Algebra in the Real World

38 Enrichment Lessons for Algebra 2

LeRoy C. Dalton

The author and publisher gratefully acknowledge Diane L. Hatchell, Ph.D., Professor, Departments of Ophthalmology and Physiology, Medical College of Wisconsin and Chief, Electron Microscope Laboratory, Wood Veterans Administration Medical Center, Milwaukee, Wisconsin for permission to use the photograph showing the endothelium layer of cells from a human cornea.

Reactor: Vincent F. Brunner
Project Editor: Elaine C. Murphy
Production Coordinator: Ruth Cottrell
Illustrator: Pat Rogondino
Cover Designer: Michael Rogondino

ISBN 0-86651-121-0
Order Number DS01339
11 12 13 14 15 – ML – 04 03 02 01 00

DALE
SEYMOUR
PUBLICATIONS

CONTENTS

PREFACE

Most mathematics teachers see beauty in pure mathematics. We like to see interrelationships of mathematical concepts and we like to see unifying ideas that tie various concepts together. Seeing beauty in the interrelationships of mathematical concepts has intrigued me ever since my undergraduate days. And seeing the unifying ideas in mathematics became a real part of my mathematics teaching during the "modern math" era when we learned to emphasize such unifying concepts as *set, relation,* and *function.* In the last fifteen years I have grown to more fully appreciate the beauty of mathematics by observing its applicability to the structure of the biological and physical worlds. This most recent interest has led me to look more enthusiastically for applications of mathematics in the worlds of recreation, work, business, and industry.

An awareness of the applications of mathematics does not come automatically to students. I have come to believe that, except for the brightest students, students do not transfer what they have learned in mathematics class to the world around them. If we feel that an understanding of the usefulness of mathematics is important, we must teach it to students by transferring the mathematics they have learned to applied areas and by training them through our own example—showing them how to look for applications.

My increased desire to find applications, from my own personal interests and from my needs as a mathematics teacher, has led me to subscribe to a variety of mathematical and scientific journals, to study books devoted to mathematics in the biological and physical worlds, and to be alert for information appearing in other types of magazines and newspapers. I now take nature walks into the woods armed with books on wildflowers and trees and I count the petals of flower blossoms, observing how often the count will be a Fibonacci number and looking for five-petal blossoms which exhibit many examples of the Golden Ratio. I collect seashells to observe the geometric designs in the colorations of their exteriors and the logarithmic spiral and helix shapes in their structures. I have found applications in the music of pianos and guitars; in the growth and decay of bacteria, heat, light, sound, and radioactive substances; in the fruits of sunflowers, pineapples, and apples; in seashells; in pine cones; in horns, beaks, teeth, fangs, claws, and tusks of certain animals; in spider webs; in cochlea of the human ear; in galaxies of stars; in the arrangement of leaves on trees; in honeycombs; in the cornea of human, cat, and rabbit eyes. The list goes on. I am hooked. I look everywhere. In fact, I have come to believe that all the biological and physical phenomena of the universe have mathematical structure; humankind has only to discover them.

I am pleased that Dale Seymour Publications has made it possible for me to share these applications with you. I am indebted to Mr. Dale Seymour for his idea to publish my transparencies after his unexpected appearance at a talk I gave for the NCTM. Also, I am fortunate to have had Ms. Elaine Murphy, an outstanding mathematics editor, as the editor for this publication. She has a real feeling for and appreciation of my applications and has done a superb job of sustaining the meaning I had intended for each application, as well as improving the manuscript through her ingenious editorial innovations. I am also indebted to my longtime friend and textbook coauthor, Mr. Vincent F. Brunner, who consented to review my manuscript and give of his wisdom and knowledge—giving ideas that have improved my manuscript.

I have had a great deal of personal fun and enjoyment in looking for applications of mathematics wherever they might occur. I have derived even more satisfaction from sharing what I have learned with my students, colleagues, and teachers around the country. This book is my opportunity to share my work with you. I hope you will find them useful as supplements for enriching your algebra classes. With a little pre-planning you will find you can match most of the applications to your course outline. And I sincerely hope that your efforts in sharing these ideas with your students will be as rewarding as mine have been.

L. C. Dalton
September, 1982

INTRODUCTION

Most mathematics teachers will agree that mathematics should not be learned in isolation from its applications. The enormous versatility and power of mathematics comes out when it is presented in a variety of contexts. A large part of the biological and physical universe has mathematical structure. The bee's honeycomb tessellates the plane. Sounds from the pipes of an organ are directly related to the size of the pipes. "Mathematics is all around us," it has been said. "We only need to look." But students must *learn to look* before they *see*. Even students who display a high level of mathematical skill often fail to see any relationship in what they are doing to real-life situations. One role of mathematics teachers is to guide students through applications of mathematics, showing them how to interpret and analyze situations from a mathematical point of view. In short, mathematics teachers must teach students to look.

To study mathematics is, in part, to see relationships and to find unifying ideas. Many a mathematician has marvelled at the connections between the binomial theorem, Pascal's Triangle, and Fibonacci numbers. The power of functions for describing real-life phenomena cannot be disputed. For the study of mathematical applications to be truly significant, they should not be treated as isolated examples. Mathematics teachers must tie them together, making the structural unity and interrelationships of the whole show through.

Many algebra students have reached a level of maturity and basic mathematical understanding that enables them to work with sequences of interesting and worthwhile problems. The job of mathematics teachers is to find significant applications that are accessible to their students and share them in a way that elucidates the mathematical structure, emphasizes the interconnections, allows students to make their own discoveries, takes students in interesting and important directions, and motivates students to ask more questions on their own. Even the best prepared, most competent and dedicated teachers need a little help. The demands on their time are simply too great.

Algebra in the Real World is a collection of mathematical applications designed to meet the needs of high school Algebra 2 teachers. The applications have been chosen especially for their accessibility to high school algebra students. The mathematics is neither trivial nor threatening; it draws from today's standard Algebra 2 curriculum—including some geometry and elementary combinatorics, quadratic and exponential relations, rational polynomial expressions, and trigonometry. The applications are chosen from familiar and appealing contexts; the lessons on computer secrets speak to an area of current concern and intense research; the section on music gives students a new look at a favorite subject; the mathematics of flowers shows up in several lessons. Each section of *Algebra in the Real World* is organized about the connections that exist between various mathematical applications. For example, the section on the Golden Ratio shows the diversity of that number's occurrences. *Algebra in the Real World* brings together a wide range of mathematical applications in a way that aids teachers in setting up a classroom environment—one in which mathematics as a creative activity will flourish.

Features of the Book

FOLLOW-ALONG
LESSONS

There are three main parts to each lesson in *Algebra in the Real World:* discussions, worksheets, and follow-up exercises. The discussions develop the content of the lesson. They are written in a narrative form—as a teacher speaking to students. Since people remember best what they discover for themselves, a number of key ideas are first presented as questions. Worksheet pages are an integral part of each lesson. They have been designed to illustrate the lesson and help enlist students' participation. Reduced versions of the worksheets appear on the discussion pages, to show how they follow along with the lesson and to help you keep your place. At the end of each lesson is a set of exercises that allow you to illuminate points in the lesson and to extend the lesson with new ideas. Answers are included. Masters for the worksheets and exercises make up the second part of the book. The worksheet masters are big and bold; you can make transparencies that will be seen—even from the back of the room. The exercise masters are designed as a handout, for in-class work or homework.

SECTIONAL THEMES

The lessons from *Algebra in the Real World* are grouped into seven different sections. Each of these sections can be studied independently of the others. It does not matter much which one is studied first and which later. You can choose the sections you wish to use based on student interest and available time. Each section is organized around a common theme—computers, music, *e*. The lessons are held together by the theme and a thread of unifying concepts and interrelationships that come out of the theme. The lessons are sequenced, building on one another as they go. The mathematics content of each lesson is indicated at the beginning of the lesson. You can judge the appropriateness of a lesson for your classes and select lessons according to students' needs.

REFERENCES

Many of the best ideas in this book grew out of material that has already been presented in one form or another. At the back of this book is a list of references keyed to sections. These references tell you where the ideas came from and provide general sources that give a context for the content. The references are particularly helpful as a starting point for further research.

USE OF CALCULATORS

Calculators and microcomputers have opened new doors in mathematics for students. Many significant applications that were unsuitable for high school classes because of their detailed calculations can now be included. Other ideas that were a part of the curriculum mainly for their computational value have either disappeared or taken on a different emphasis. Logarithms are one example. You will find that the lessons in this book may be easier to handle using calculators. A number of lessons use data that can be obtained using calculators. If they are available, you will be able to give your students a greater chance to be involved in the generation of that data.

All the numerical approximations in this book were checked against a *Texas Instruments* scientific calculator, the TI-35. Other calculators may give slightly different answers depending on the number of digits they can hold and the way their calculations are made. Be prepared for small variations in values if your tools are not the same as the TI-35.

VARIED USES   *Algebra in the Real World* was put together to give you a chance to supplement and enrich topics you've worked on in your algebra classes. The lessons should help you show students some real uses of what they've learned. You may want to use individual lessons or complete sequences of lessons for in-class presentations. You could make transparencies of the worksheet masters to allow you to work with the whole class. You may also find that the material in this book can be used in math clubs or student math conferences. Either teachers or students could present the ideas. The applications in *Algebra in the Real World* are ideally suited for use in inservice workshops and teacher math conferences as well.

Beyond this Book

The applications in *Algebra in the Real World* are only a beginning. Once you've seen a few applications like these, you want to find more. You will begin to see applications all around you— on nature walks, in big cities, in your dentist's office. Mathematical and scientific journals are excellent sources for ideas as are books and magazines devoted to the biological and physical worlds. The daily newspaper is another starting point for ideas. Your imagination is the only limit on the applications you can discover. And you will find that applications you put together on your own are often the ones that will excite you (and therefore your students) the most.

ALGEBRA IN
THE REAL WORLD

DISCUSSIONS

ONE

PROTECTING A COMPUTER'S SECRETS*

The growing use of computers has created serious problems with protecting confidential information. Today, people use computers to store all kinds of information. Computer data banks may contain medical records, credit ratings, contract information, or secret formulas among other things. And whatever goes into a computer can come out.

Problems called "hard" problems may someday be used to help protect a computer's secrets. (In computer jargon these problems are called nondeterministic, polynomial-time problems, or just NP problems.) These "hard" problems are very difficult to solve because the computer could take years, or sometimes even centuries to do all the necessary calculations. Why do these problems take so long to solve? The computer must test every possible case to arrive at a solution. And, as the number of items in the problem increases, the number of possible cases to test literally explodes.

Finding the Shortest Route and *Bin-Packing* describe two different "hard" problems that could be used to protect a computer's secrets. Distorting Data describes another kind of problem that could be used to foil a computer thief.

*This idea and the problems that follow are based on information from the following articles.

"Trial-and-error game that puzzles fast computers" by Gina Bari Kolata appeared in the October 1979 issue of *The Smithsonian*, pages 90–96.

"The Mathematics of Public-Key Cryptography" by Martin E. Hellman appeared in the August 1979 issue of *Scientific American*, pages 146–157.

"Some finite numbers are pretty big, too" by William F. Allman in the June issue of *Science '82*, page 63.

"Math of a Salesman" in the November issue of *Science '82*, pages 7–8.

FINDING THE
SHORTEST ROUTE

MATHEMATICS CONTENT:
Permutations,
Factorial Notation,
Scientific Notation

One type of "hard" problem is the so-called Traveling Salesman Problem. Here's one way of stating it.

> A sales representative, call her Ms. Johnson, must visit a certain number of cities, each exactly once. What is the shortest possible route she can take?

Solving the problem is simple if there are only a few cities on the route. Look at page 1. The first diagram models the problem for two cities, *A* and *B*. The diagram shows that there are only two routes to test, route *AB* and route *BA*. They are obviously the same length. (One way to find the number of routes is by counting the permutations of two items taken two at a time. You get 2! or 2.)

If Ms. Johnson has three cities to visit, there are six different routes she can take. In this diagram the routes are *ABC*, *ACB*, *BAC*, *BCA*, *CAB*, and *CBA*. (Six is the same as 3!, the number of permutations of three items taken three at a time.)

How many possible routes are there if Ms. Johnson visits four cities? You can make a list of the routes and count them. (There are 4! or 24 possible routes, the number of permutations of four items taken four at a time.)

Suppose Ms. Johnson visits five cities? Look at page 2. One possible route on the five-city diagram might be *ADCEB*. Can you suggest some others? How many routes are there altogether? The first few cases reveal a pattern. (It suggests that the number of routes is 5! or 120, the permutations of five items taken five at a time.)

So far, the problem is simple. But look at what happens if the sales representative visits 10 cities. The diagram shows a model of the problem. One sample route is *ABEDCFIGHJ*. By using the pattern that has been established, you can

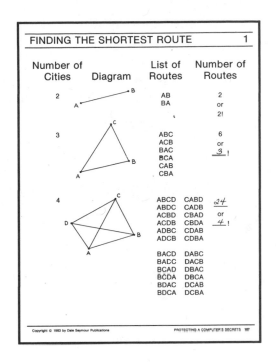

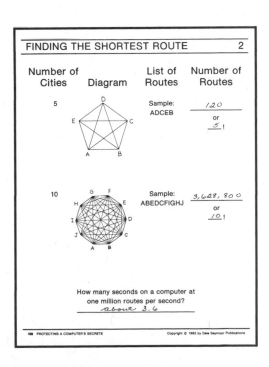

count the number of possible routes. (There are 10! or 3,628,800 possible routes—a lot of testing without a computer.)

Suppose that a computer can test one million routes per second. How long would it take the computer to test 10! routes? (About 3.6 seconds. No sweat for the computer.)

Think about 18 cities. Can you complete the statements on page 3? There are 18! or about 6.4024×10^{15} routes. How well will the computer do now? (A fast computer testing one million routes per second would require 203 years to test all of these routes.)

Now try 60 cities—a more realistic problem. How many routes must a computer test? How long would it take the computer to test them at one million routes per second? (Billions of centuries! That's why computer experts call this problem a "hard" problem.)

Just for fun, I found out how long it would take the computer to check all 60 routes. (This is a good chance to test your ability to handle scientific notation.) Here's how. (Follow along on page 4.)

1. Find the number of routes.
2. Find the number of seconds in one year.
3. Divide the number of routes by both the number of seconds in a year and the number of routes the computer can check in a second. (The given calculation assumes the computer checks one million routes per second.)

Needless to say, if a computer had to find the shortest route for visiting each of sixty cities exactly once by checking all of the routes, it would be busy for a long long time. If a thief had to solve this problem to find a company's secret data, it would never be found. The computer time needed to solve this problem is 2.639×10^{66} *lifetimes* of 100 years each!

FINDING THE SHORTEST ROUTE   3

For 18 cities there are __18__! or about __6.4024__ $\times 10^{15}$ routes.

At one million routes per second, it would take a computer about __203__ years to test all of these routes.

For 60 cities, there are __60__! routes. At one million routes per second it would take

   a. years.
   b. decades.
   c. centuries.
   d. billions of centuries.

Copyright © 1983 by Dale Seymour Publications   PROTECTING A COMPUTER'S SECRETS 109

FINDING THE SHORTEST ROUTE   4

How much time for 60 cities?
1. Find the number of routes.
   $60! \approx$ __8.321__ $\times 10^{81}$

2. Find the number of seconds in a year.

$$\frac{3600}{\text{seconds in one hour}} \times \frac{24}{\text{hours in one day}} \times \frac{365}{\text{days in one year}} = \frac{31,536,000}{\text{seconds in one year}}$$

3. Divide the number of routes by both the number of seconds in a year and the number of routes the computer can check in a second.

$$\frac{8.32 \times 10^{81} \text{ routes}}{31,536,000 \frac{\text{seconds}}{\text{year}} \times 1,000,000 \frac{\text{routes}}{\text{second}}}$$

$= 2.639 \times 10^{68}$ years

$= 2.639 \times 10^{66}$ centuries

$= 2639 \times 10^{63}$ centuries

$= 2639$ vigintillion centuries

110 PROTECTING A COMPUTER'S SECRETS   Copyright © 1983 by Dale Seymour Publications

EXERCISES

1. Find the number of routes that pass once and only once through each of 25 known cities.

2. Suppose a sales representative, Mr. Calling, knows the distance between each pair of the 25 cities described in Exercise 1. He wishes to determine the shortest route. Find the time needed for a computer to check each complete route if it can check one million routes per second.

3. Suppose that in one zone a telephone company has 50 pay phones from which coins must be collected periodically. The collection supervisor who picks up the coins wishes to find the shortest route to follow when visiting each pay phone on the route exactly once.

   a. Find the number of routes to be tested, and
   b. Determine the computer time needed to check the length of each route. (Assume the computer can check one million routes per second.)

ANSWERS

1. 25! or about 1.5511×10^{25} routes

2. about 4.9186×10^{11} years

3a. 50! or about 3.0414×10^{64} routes

3b. about 9.6442×10^{50} years

BIN-PACKING

In how many different ways can you pack *n* objects, say Ping-Pong balls, into two bins of unlimited size? This doesn't sound like a very practical problem, but it is—when it comes to computers. Bin-packing problems are "hard" problems, too.

When there is only one Ping-Pong ball, there are only two different ways to pack the bins. (See page 1.) The ball can either be in the first bin or the second bin. If you add another Ping-Pong ball, the number of possibilities doubles. The diagram shows all the different ways the two bins can be filled with two balls. (There are 2^2 or 4 different ways.)

The number of possibilities doubles again if you try to pack three Ping-Pong balls into the two bins. (See page 2.) All three balls could go in Bin 1 or all three could go in Bin 2. Can you show the other possibilities? (In all, there are 2^3 or 8 possible packings.)

If you organize the data you collect about the number of ways to pack Ping-Pong balls into two bins, you'll see a pattern develop. (See page 3.) The number of ways you can pack balls into two bins of indefinite size is a function of the number of balls you use. The function is an *exponential function, $f(x) = 2^x$* where *x* represents the number of balls.

Now, suppose that a computer can look at one million different packings in one second. It will take that computer about 1/1000 second to check all the different packings of 10 Ping-Pong balls into the two bins and it will take about 1 second to check for 20 balls. The computer time increases considerably for 30 balls (18 minutes). What about for 100 balls? It will take the computer centuries! (About 400,000 billion centuries!)

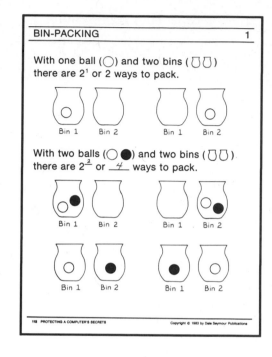

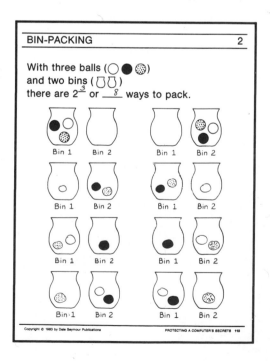

If finding the key to a computer's confidential information required checking every single packing of 100 balls into 2 bins, then the key and the information would be safe. It would take a thief longer than the universe is old to find it.

EXERCISES

1. Draw a diagram showing all 16 different ways in which four Ping-Pong balls can be packed into two bins. (To check yourself as you work, it's helpful to know how many different ways n items can be chosen r at a time. A formula for finding this answer is
$$_nC_r = \frac{[n(n-1)(n-2) \cdots (n-r+1)]}{r!}$$

2. Draw a diagram showing all 32 different ways in which five Ping-Pong balls can be packed into two bins.

3. Find the number of different ways fifty Ping-Pong balls can be packed into two bins. Write your answer in scientific notation.

4. Find the approximate computer time required to test all the packings of fifty Ping-Pong balls into two bins using a computer that tests one million packings per second.

ANSWERS

1. Answers will vary.

2. Answers will vary.

3. about 1.1259×10^{15}

4. about 36 years

DISTORTING DATA

A knot is often very easy to tie but, if you don't know the trick, it's not very easy to untie. So it is with certain mathematical functions. They can transform data into an almost irretrievable form. These functions turn information into an unintelligible mess, unless you know the trick to them.

For example, suppose the numbers 1, 2, 3, 4, 5, and 6 are transformed into 2, 7, 12, 17, 22, and 27. With a little thought, it's not too difficult to figure out how the numbers were changed. Can you guess? Try graphing to get an idea. (Take any number from the first set, multiply by 5 and subtract 3. Voila! You get a corresponding number from the second set. This transformation can be described by the functional equation $f(x) = 5x - 3$.) It's a *linear function*. The graph show the points lined up. Easy, you say.

Try another transformation. Suppose the numbers 2, 7, 12, 17, 22, and 27 are now changed to 2, 0, 5, 3, 1, and 6. Can you figure out how the numbers were changed this time? I doubt it. Even a graph isn't very helpful. (The points are scattered. They don't follow any well-known function pattern.)

There really is a function that transforms the data this way. Here's the trick. Take any number from the first set and divide by 7. Look at the remainder. This kind of transformation can be described by a *modular function*. In our case, mathematicians would use the equation $g(x) = f(x)$ mod 7. It says that the values of g are the remainders you get when you divide values of f by 7. Try the rule on all the numbers in the transformation to check. Modular functions make good disguises, don't you think?

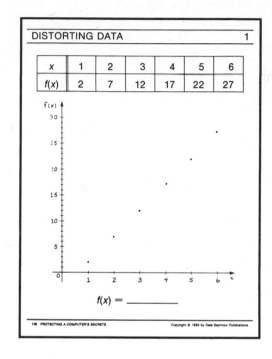

DISTORTING DATA   1

x	1	2	3	4	5	6
f(x)	2	7	12	17	22	27

$f(x) =$ _____

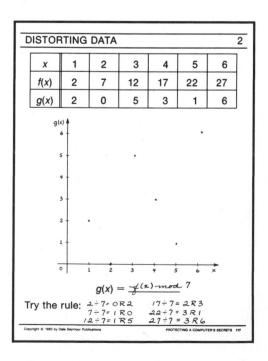

DISTORTING DATA   2

x	1	2	3	4	5	6
f(x)	2	7	12	17	22	27
g(x)	2	0	5	3	1	6

$g(x) = f(x) \bmod 7$

Try the rule:
2 ÷ 7 = 0 R 2   17 ÷ 7 = 2 R 3
7 ÷ 7 = 1 R 0   22 ÷ 7 = 3 R 1
12 ÷ 7 = 1 R 5   27 ÷ 7 = 3 R 6

EXERCISES

1. Find an equation of the form $f(x) = mx + b$ for the linear function that describes the following transformation.

x	1	2	3	4	5	6
$f(x)$	7	10	13	16	19	22

2. Find an equation of the form $h(x) = ax^2 + b$ for the quadratic function that describes the following transformation.

x	1	2	3	4	5	6
$h(x)$	-4	2	12	26	44	66

3. Suppose $g(x) = f(x) \bmod 6$ and $f(x) = 5x - 4$. Complete the following table.

x	1	2	3	4	5	6	7
$f(x)$	1	6	11		21		
$g(x)$	1	0		4			

4. Suppose $g(x) = f(x) \bmod 9$ and $f(x) = 2x^3$. Complete the following table.

x	1	2	3	4	5	6
$f(x)$	2	16	54		250	
$g(x)$	2	7		2	7	

5. In 1982, several mathematicians devised a rapid solution to an NP problem. What was the problem and how long did it take to solve?

ANSWERS

1. $f(x) = 3x + 4$

2. $h(x) = 2x^2 - 6$

3. $f(4) = 16, f(6) = 26,$ $f(7) = 31; g(3) = 5, g(5) = 3,$ $g(6) = 2, g(7) = 1$

4. $f(4) = 128, f(6) = 432;$ $g(3) = 0, g(6) = 0$

5. The mathematicians showed that a 97-digit number they were given was, in fact, a prime number. It took their computer 77 seconds to solve it.

TWO

PACKING PROBLEMS*

Believe it or not, mathematicians solve problems that are practical. But, as often as not, a mathematician works on a problem that appears to have no practical significance. Why? A famous mountain climber gave the answer—"because it's there."

There is a practical side to impractical problems. Who knows, there may be a real use for them some day. Some of the most "useless" results have found a use. For example, finding out how many different ways marbles can be packed into infinite spaces doesn't sound so terribly practical. It's a problem that has found a possible application, though—hiding computer secrets. Packing problems dealing with *finite* spaces have a great potential for practicality. What's the best way to fit certain items into a given space? (What's the greatest number of tuna cans you can put into a given carton?) Perhaps you'll be the one to find an application for the following packing problems.

*The ideas that follow are based on ideas from:

"The diverse pleasures of circles that are tangent to one another" in the Mathematical Games section of the January 1979 issue of *Scientific American*, pages 18–28.

"Close Packing of Equal Spheres" in *Introduction to Geometry* by H. S. M. Coxeter, pages 405–407.

The Teacher's Guide to the *Math in Nature Posters* of Creative Publications written by Alan Hoffer, page 5.

"The Packing of Equal Circles in a Square" by Michael Goldberg in the January 1970 issue of *Mathematics Magazine*, pages 24–30.

FITTING CIRCLES INTO SQUARES

Suppose you have a square that measures one unit on a side. What's the biggest circle you can fit into that square? What if you want to fit two identical circles into the square? How big can they be? What about for three identical circles? What about for 4, or more?

A mathematician named Goldberg published some of the answers to these questions. The diagram shows his answers for the first nine cases. For example, the largest two circles you can fit into a one-unit square have diameters slightly more than 0.58 units. Goldberg also gave his best guesses for up to 27 circles and some guesses for even greater numbers. But he couldn't prove them. So far, no one has.

Take another look at the diagram. Notice that some of the packings "fill up" the squares better than others. There's a way to actually measure how well a particular packing of circles fills up a square. Add up the areas of the circles in a square. Divide your answer by the area of the square itself. This number is called the *density* of the packing. To find the density for five circles in a square, for example, use 5 for n, 0.41/2 for r, and 3.14 for π in the formula given. The density is about 0.6598. What's the density for 6 circles? (0.6448)

Because the circles never quite fill up an entire square, the density of each packing will always be a number less than one. The best packings will have the greatest densities (densities closest to 1). Which of the packings in the diagram do you think is best?

As the number of circles packed in the square increases, there doesn't seem to be any pattern to how the densities change. The numbers "bounce" up and down. It happens, however, that as the number of circles increases, the densities approach 0.9069. (The limit obtained by the familiar close packing of circles with their centers on a regular lattice of equilateral triangles.)

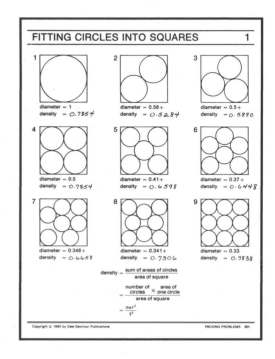

FITTING CIRCLES INTO SQUARES   1

1 diameter = 1   density = 0.7854
2 diameter = 0.58+   density = 0.5284
3 diameter = 0.5+   density = 0.5890
4 diameter = 0.5   density = 0.7854
5 diameter = 0.41+   density = 0.6598
6 diameter = 0.37+   density = 0.6448
7 diameter = 0.348+   density = 0.6658
8 diameter = 0.341+   density = 0.7306
9 diameter = 0.33   density = 0.7838

$$\text{density} = \frac{\text{sum of areas of circles}}{\text{area of square}}$$

$$= \frac{\text{number of} \atop \text{circles}} {\times} {\text{area of} \atop \text{one circle}} \over {\text{area of square}}$$

$$= \frac{n\pi r^2}{1^2}$$

EXERCISES	ANSWERS

EXERCISES

1. Find the density of the packing of 3 circles into a unit square (the packing as described in this article). Use 3.1415927 for π.

2. Find the density for the 4-circle packing.

3. Find the density for the 7-circle packing.

4. Find the density for the 8-circle packing.

5. Find the density for the 9-circle packing.

6. Goldberg had to decide how to best pack the circles into the square to come up with his figures. Find out how he suggested the circles should be packed. (The answer can be found in the January, 1979 *Scientific American* or in the January, 1970 issue of *Mathematics Magazine*.)

ANSWERS

1. 0.5890

2. 0.7854

3. 0.6658

4. 0.7306

5. 0.7854

6. There must be a structure of touching circles that connects all 4 sides of the square and such that each circle touches at least 3 other circles or a side of the square.

THE BEE SOLVED
A PACKING PROBLEM

MATHEMATICS CONTENT:
Ratio, Radicals, Rational Expressions,
Area of Circle, Properties of
30°-60° Right Triangle, Equilateral Triangle,
and Regular Hexagon

A long time ago the bee figured out how to store its honey. The bee chose the most efficient and economically-shaped packing container for the occasion, a regular hexagonal prism. Regular hexagonal prisms fit together without overlapping so that their top surfaces completely cover the plane, forming the bee's familiar honeycomb structure. Mathematicians say that the top surfaces, the hexagons, *tessellate* the plane. Only three regular polygons can tessellate the plane—the regular hexagon, the square, and the equilateral triangle. The bee had only three reasonable choices.

To show that the bee's regular hexagon is the most efficient and economical choice, you must calculate the *densities* of the three tessellations and compare them. The densities will give you a measure of how well the bee's honey deposit fills the space allowed. To find the densities, you inscribe a circle into each polygon. (The circle represents the bee's honey deposit.) Then you find the ratio of the area of the inscribed circle to the area of the polygon.

The chart gives values for the densities of each tessellation. The decimal approximation for the density of a triangular tessellation is given. What are decimal approximations for the other two tessellations? (If you don't have a scientific calculator, use 3.14 for π and 1.732 for $\sqrt{3}$.) Which tessellation gives the densest packing? (The hexagonal tessellation gives the densest packing, making the bee look pretty good.)

Actually calculating the densities for each of the tessellations requires some elementary algebra and a few well-known properties of triangles. The second page points out the calculations you need to make to determine the density of a triangular tessellation. Knowing that the altitudes, medians, and angle bisectors of an equilateral triangle are concurrent (pass through the same point) and that their common point is two-thirds the distance from each vertex to the midpoint of the opposite side allows you to make the calculations.

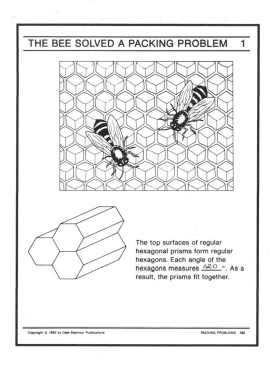

THE BEE SOLVED A PACKING PROBLEM   1

The top surfaces of regular hexagonal prisms form regular hexagons. Each angle of the hexagons measures _120_ °. As a result, the prisms fit together.

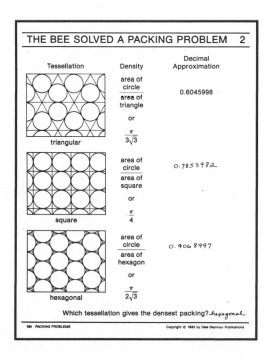

THE BEE SOLVED A PACKING PROBLEM   2

Tessellation	Density	Decimal Approximation
triangular	$\dfrac{\text{area of circle}}{\text{area of triangle}}$ or $\dfrac{\pi}{3\sqrt{3}}$	0.6045998
square	$\dfrac{\text{area of circle}}{\text{area of square}}$ or $\dfrac{\pi}{4}$	0.7853982
hexagonal	$\dfrac{\text{area of circle}}{\text{area of hexagon}}$ or $\dfrac{\pi}{2\sqrt{3}}$	0.9068997

Which tessellation gives the densest packing? *hexagonal*

Here's what you do.

1. Find the number of degrees in each angle of the small triangle.

2. Determine the lengths of the sides of the small triangle (in terms of r).

3. Find the length of one side of the large equilateral triangle (in terms of r).

4. Find an expression for the area of the equilateral triangle in terms of its sides.

Putting the information in steps 1–4 together gives you the following expression for the density.

$$\frac{\pi r^2}{\dfrac{(2r\sqrt{3})^2\sqrt{3}}{4}}$$

This expression simplifies to $\pi/(3\sqrt{3})$. Can you show how it's done?

The easiest calculation is for the density of a square tessellation. (The area of the inscribed circle is πr^2 and the area of the square is $4r^2$ because each side of the square is $2r$. The density, then, is $(\pi r^2)/(4r^2)$ or $\pi/4$.

Finding the density of a regular hexagonal tessellation is much like finding the density for the triangular tessellation; both calculations use the formula for the area of an equilateral triangle. In the case of the hexagon, you use six times the area of an equilateral triangle. Why? (A regular hexagon is composed of six equilateral triangles. The diagram shows them.) In order to calculate the area of the equilateral triangles, you must first find the length of a side in terms of r. A 30°-60° right triangle comes to the rescue. In terms of r, the hypotenuse of that small triangle shown in the diagram is $2r/\sqrt{3}$ and the shortest side is $r/(\sqrt{3})$. The density, then, is

$$\frac{\pi r^2}{6\left(\dfrac{\left(\dfrac{2r}{\sqrt{3}}\right)^2\sqrt{3}}{4}\right)}$$

which simplifies to $\pi/(2\sqrt{3})$.

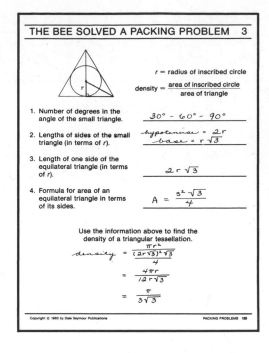

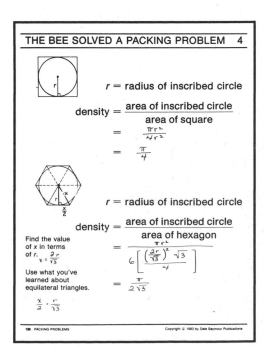

Mathematics agrees with the bee. A regular hexagonal tessellation gives the densest packing.

Nature's hexagonal tessellations are not exclusive to the bee. There are hexagonal tessellations in your own eye! Page 5 shows a magnified version of a layer of the human cornea (the endothelium, or most posterior layer, of the cornea). Do you see the hexagonal cells that cover the plane of the photograph? Researchers have found this same kind of tessellation in the endothelium layer of the corneas of cats and rabbits, too.

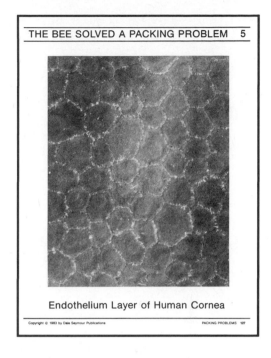

THE BEE SOLVED A PACKING PROBLEM   5

Endothelium Layer of Human Cornea

EXERCISES

1. Explain why a regular pentagon cannot tessellate the plane.

2. Explain why a regular octagon cannot tessellate the plane.

ANSWERS

1. If you fit three pentagons together at a vertex, they will form a 324° angle—leaving a 36° "space." It is impossible to fit another pentagon into that space if the pentagons must remain in the plane. In other words, *no* multiple of the vertex angles equals 360°.

2. If you fit two octagons together at a vertex, they will form a 270° angle—leaving a 90° "space." It is impossible to fit another octagon into that space if the octagons must remain in the plane. In other words, *no* multiple of the vertex angles equals 360°.

3. Explain why the regular hexagon, the square and the equilateral triangle are the only three regular polygons that can tessellate the plane.

4. There are many *non-regular* pentagons that will tessellate the plane. Give one example.

3. These are the only regular polygons for which a multiple of the vertex angles equals 360°.

4. Answers will vary. Two examples are as follows.

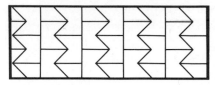

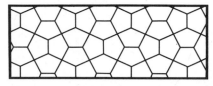

For more information about this topic see Schattschneider, Doris, "Tiling the Plane with Congruent Pentagons," *Mathematics Magazine,* Vol. 51, No. 1, January, 1978, pp. 29–44.

A VERY SPECIAL NUMBER CALLED e*

One of the most interesting numbers in mathematics is e. Like π, it is *irrational* (cannot be expressed as the quotient of two integers) and *transcendental* (cannot be expressed as a solution to a polynomial equation having rational coefficients). It is the most used base (as in base and exponent) in calculus—the number whose natural logarithm equals 1. Its decimal expansion, correct to ten decimal places, is 2.7182818285 . . . .

Just as π shows up in a variety of mathematical situations, e makes many appearances. People have found it in a wide assortment of relationships from mere curiosities to very remarkable and elegant results. The following lessons present only a few of e's appearances, but perhaps they will point out why e is the second most famous transcendental number—second only to π.

*The lessons that follow are based on ideas from the following.

"Applications of Mathematics in the new Spirit of St. Louis Arch," by Lee E. Boyer.

Bakst, "Mathematics: Its Magic and Mastery," pages 313–328.

"Is Exponentiation Commutative?" by Warren B. Manhard, 2nd in the January 1981 issue of the *Mathematics Teacher*, pages 56–60.

"In some patterns of numbers or words there may be less than meets the eye," in the Mathematical Games section of the September 1979 issue of *Scientific American*, pages 22–25.

"The imaginableness of imaginary numbers," in the Mathematical Games section of the August 1979 issue of *Scientific American*, pages 18–24.

THE FAMOUS ST. LOUIS CATENARY

MATHEMATICS CONTENT:
Exponential Functions,
e

Historically, St. Louis, Missouri was an important starting point for pioneers on their way West. To commemorate its role as the gateway to the West, the city of St. Louis built a 630-foot stainless steel structure called the Gateway Arch. The design and construction of this imposing arch was an incredible feat of engineering.

The Gateway Arch is famous not only for its symbolic role as the gateway to the West, but also for what it represents mathematically. It takes the shape of a *catenary* curve. (If you hold a chain at its two ends, letting the chain hang freely, you'll form another catenary curve.) In architecture, the catenary shape makes an extremely stable arch because the force of the arch's weight acts along its legs directly into the ground.

A general catenary curve can be described by the following equation.

$$y = \frac{a}{2}\left(e^{\frac{x}{a}} + e^{-\left(\frac{x}{a}\right)}\right)$$

where a is a real nonzero constant and x is real

To see how values of the constant a affect the shape of the curve, sketch several different catenary curves on the same set of axes. Try $a = 2$, $a = 4$, and $a = 6$. The graph on page 2 shows a catenary curve for $a = 4$. The curve was determined by finding points for all integer values of x from -7 to 7 and using 2.71828 as an approximation for e.

Notice how the graphs differ for different values of a. The values of a determine the y-intercepts of the curves. Do you see why? The values also affect the degree of curvature (the slope) of the curves. As the values of a increase, do the slopes decrease (become more gradual) or increase? (They decrease.)

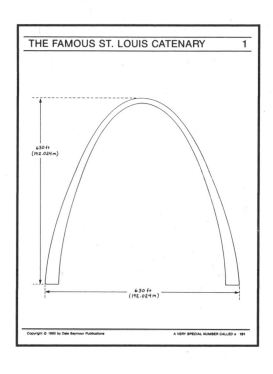

THE FAMOUS ST. LOUIS CATENARY   1

630 ft
(192.024 m)

630 ft
(192.024 m)

Copyright © 1983 by Dale Seymour Publications   A VERY SPECIAL NUMBER CALLED e   191

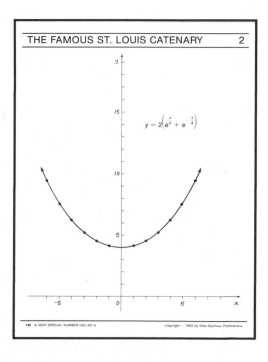

THE FAMOUS ST. LOUIS CATENARY   2

$$y = 2\left(e^{\frac{x}{4}} + e^{-\frac{x}{4}}\right)$$

132 A VERY SPECIAL NUMBER CALLED e   Copyright · 1983 by Dale Seymour Publications

The Gateway Arch actually is an inverted catenary; it's upside-down. Its height from the very highest point to the ground is 630 feet (192.024 meters). Its width from one leg to the other along the ground is also 630 feet (192.024 meters).

EXERCISES

1. Sketch the graph of a catenary curve for which $a = 1$. Find points for integer values of x from 0 to 4. Also find points for $x = 3.25$, $x = 3.5$, and $x = 3.75$. Use 2.718 for e.

2. Explain how the shapes of catenaries with $a = 2$ and $a = 1/2$ differ.

3. Explain how the graphs of catenaries with *positive* values of a differ from the graphs of catenaries having *negative* values of a.

4. Give an equation for the catenary curve that has the same shape as $y = e^{x/2} + e^{-(x/2)}$ and opens in the same direction but has a *y*-intercept of 4.

ANSWERS

1. The curve opens up and has a *y*-intercept of 1. The following points are on the curve: (0, 1), (1, 1.543), (2, 3.762), (3, 10.068), (3.25, 12.915), (3.5, 16.573), (3.75, 21.27), (4, 27.308)

2. The catenary for which $a = 1/2$ has a minimum point (at the *y*-intercept) closer to zero than the catenary for which $a = 2$. Also, when $a = 1/2$, the curve opens very slowly (has a steeper slope).

3. Catenaries having *positive* values for a open up; those having *negative* values for a open down.

4. $y = e^{x/2} + e^{-(x/2)} + 2$

GROWTH AND DECAY

When it comes to growth and decay, very few functions are as useful as the exponential function $y = ne^{kt}$. So many phenomena in the physical and biological worlds can be described by this function that often it is called the growth and decay formula. In the formula, y stands for the final amount of an item, n stands for the initial (starting) amount, k is a growth or decay constant, and the variable t stands for time elapsed. The constant k is greater than 0 when the function describes growth and it is less than 0 when the function describes decay. The value of the constant e is approximately 2.72.

The list on page 1 points out only a few of the many situations that can be described using the growth and decay formula. Can you find more?

A typical bacterial growth problem can be stated in the following way.

> For a certain strain of bacteria, the growth constant k is 0.867 when t is measured in hours. How long will it take for 20 bacteria to increase in number to 500 bacteria?

In this problem, the final number of bacteria, y, is 500 and the initial number of bacteria, n, is 20. You are given a value for k and you know an approximation for e, 2.72. The unknown quantity is t, the amount of time. Page 2 shows the equation you must solve.

If you think calculators and computers have made logarithms old-fashioned, think again. Logarithms are excellent tools for solving exponential equations. In the problem given here, you can use logarithms to help isolate t. By taking the logarithm of both sides of the equation, you get $\log 25 = (0.867t)(\log 2.72)$. Solving for t gives the following equation.

$$t = \frac{\log 25}{(0.867)(\log 2.72)}$$

GROWTH AND DECAY 1

$y = ne^{kt}$ where $\begin{cases} y = \text{final amount} \\ n = \text{initial amount} \\ e \approx 2.72 \\ k = \text{constant} \\ t = \text{time elapsed} \end{cases}$

Practical Applications:

- Growth of bacteria
- Cooling of coffee
- Radioactive decay
- Intensity of light passing through a medium
- atmospheric pressure at various heights above sea level
- electrical conductivity of glass at various temperatures
- decay of sound

GROWTH AND DECAY 2

For a certain strain of bacteria, the growth constant k is 0.867 when t is measured in hours. How long will it take for 20 bacteria to increase in number to 500 bacteria?

$$500 = 20(2.72)^{(0.867)t}$$
$$25 = 2.72^{(0.867)t}$$
$$\log 25 = \log 2.72^{(0.867)t}$$
$$= (0.867t)(\log 2.72)$$
$$t = \frac{\log 25}{(0.867)(\log 2.72)}$$
$$\approx \frac{1.3979}{(0.867)(0.4346)}$$
$$\approx 3.710$$

It will take about 3.7 hours.

To find a decimal value for t, you can use logarithms (and a log table) or you can use a scientific calculator. Using logarithms, a table with mantissas accurate to four decimal places gives an answer of about 3.704 hours. Using a calculator gives about 3.710. A calculator and 2.71828 for e gives about 3.712. (Using natural logarithms rather than base 10 logarithms gives $t = (\ln 25)/(0.867)$ and, from a calculator, $t = 3.713$.)

<table>
<tr><td>

EXERCISES

1. Solve the bacterial growth problem for 40 bacteria growing to 600 bacteria.

2. Solve the bacterial growth problem for 15 bacteria growing to 400 bacteria.

3. Bacteria of a certain type can grow from 80 to 164 bacteria in 3 hours. Find the growth constant for this type of bacteria.

4. Tell whether the growth or decay constant is positive, or negative for each of the following applications.

   a. decay of sound
   b. decomposition of radioactive substances
   c. intensity of light passing through a medium

5. In 10 years, the radioactive mass of a 200-gram sample decays to 100 grams. (This 10-year period is called the half-life of the radioactive material.) Write an equation that describes this situation, and then find the decay constant for this radioactive substance.

</td><td>

ANSWERS

1. about 3.121 hours (using 2.72 and a calculator)

2. about 3.785 hours (using 2.72 and a calculator)

3. about 0.239 (using 2.72 for e and a calculator)

4. a. negative
   b. negative
   c. negative

5. $100 = 200e^{10k}$; k is about -0.069

</td></tr>
</table>

LOOKING FOR SOLUTIONS

Although you may never have asked yourself, "When does $x^y = y^x$," a lot of mathematicians have. Can *you* find the integer solutions to the equation? Any ordered pairs for which $x = y$ are solutions. However, $4^4 = 4^4$ and $6^6 = 6^6$ are not very exciting results. A little trial-and-error will give you two, more interesting solutions—(2, 4) and (4, 2). Are there *negative* integer solutions to the equation for which x and y are *not* equal? Yes. (The only negative integer solutions under these conditions are $(-2, -4)$ and $(-4, -2)$.)

Once you've found a positive solution to the equation, you can find a related negative solution. For example, $(-2, -4)$ is related to (2, 4). From now on, think about *positive* solutions only. The negative solutions will follow.

So far, you've only looked for the integer solutions to $x^y = y^x$. Are there other kinds of solutions and, if so, how do you find them? The graph on the first page shows the positive solutions for $x^y = y^x$. There are two parts—a straight line representing all solutions for which $x = y$, and a curve representing those solutions for which $x \neq y$. As the values of x increase, the curve gets closer and closer to $y = 1$. As the values of y increase, the curve gets closer and closer to $x = 1$. The number e makes an appearance, too. The curve crosses the line at (e, e)!

The graph should give you an idea of the kinds of solutions you can find for $x^y = y^x$, but it doesn't really give you specifics. Does $(9/4)^{27/8} = (27/8)^{9/4}$? Does $(3\sqrt{3})^{\sqrt{3}} = (\sqrt{3})^{3\sqrt{3}}$? How do you find the coordinates of specific points on the curve?

First, what about rational number solutions? According to Martin Gardner, you can find the positive rational number solutions using the following pair of equations.

$$x = \left(1 + \frac{1}{n}\right)^{n+1} \qquad y = \left(1 + \frac{1}{n}\right)^{n}$$

where $x > y$ and n is a positive integer

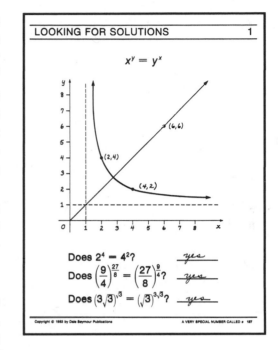

LOOKING FOR SOLUTIONS   1

$x^y = y^x$

Does $2^4 = 4^2$?   *yes*

Does $\left(\frac{9}{4}\right)^{\frac{27}{8}} = \left(\frac{27}{8}\right)^{\frac{9}{4}}$?   *yes*

Does $(3\sqrt{3})^{\sqrt{3}} = (\sqrt{3})^{3\sqrt{3}}$?   *yes*

The constant n is called a *parameter,* a constant that can be varied. (Parameters are also used in the equation of a line $y = mx + b$. In this case, the constants m and b are parameters.) If you complete the table on the top of page 2, you will obtain three different solutions to $x^y = y^x$. Where are these solutions graphed?

Some of the solutions to $x^y = y^x$ have irrational values. Warren Manhard and his high school students developed the following parametric equations. In these equations, k is a positive real number and $k \neq 1$.

$$x = k^{\frac{1}{(k-1)}}$$

$$y = k^{\frac{k}{(k-1)}}$$

These equations give many solutions to the equation $x^y = y^x$, including many irrational number solutions. Completing the table of values on page 3 using Manhard's equations will give you a solution you've seen before and several new ones. Once you've found the solutions, try placing them on the graph.

You may want to find other points on the curve using the formulas. You may also want to read the article about Manhard's work. It's an update on the progress of finding the real and complex solutions to $x^y = y^x$. The work people have done with $x^y = y^x$ shows how a problem that is quite simple (and possibly trivial) within one set of numbers becomes more complex (and often more interesting) within other sets of numbers.

LOOKING FOR SOLUTIONS 2

These equations give positive *rational* solutions to $x^y = y^x$.

$$x = \left(1 + \frac{1}{n}\right)^{n+1}$$

$$y = \left(1 + \frac{1}{n}\right)^{n}$$

where $x > y$ and n is a positive integer.

n	x	y
1	4	2
2	$27/8$	$9/4$
3	$256/81$	$64/27$
4	$3125/1024$	$625/256$

These equations give positive solutions to $x^y = y^x$, some of which are *irrational.*

$$x = k^{\frac{1}{(k-1)}}$$

$$y = k^{\frac{k}{(k-1)}}$$

where k is a positive real number and $k \neq 1$.

k	x	y
$\frac{4}{3}$	$64/27$	$256/81$
2	2	4
3	$\sqrt{3}$	$3\sqrt{3}$
4	$\sqrt[3]{4}$	$4\sqrt[3]{4}$
5	$\sqrt[4]{5}$	$5\sqrt[4]{5}$

Compare the values for $n=3$ to those for $k = \frac{4}{3}$.

A VERY SPECIAL NUMBER CALLED e Copyright © 1983 by Dale Seymour Publications

EXERCISES

1. Verify that $(-2, -4)$ and $(-4, -2)$ are solutions to the equation $x^y = y^x$.

2. Show that $(9/4, 27/8)$ is a solution to $x^y = y^x$.

ANSWERS

1. $(-2)^{-4} = 1/(-2)^4 = 1/16$;
   $(-4)^{-2} = 1/(-4)^2 = 1/16$

2. $(9/4)^{27/8} = ((3/2)^2)^{27/8} = (3/2)^{27/4}$;
   $(27/8)^{9/4} = ((3/2)^3)^{9/4} = (3/2)^{27/4}$

3. Show that $(3\sqrt{3}, \sqrt{3})$ is a solution to $x^y = y^x$.

3. Find $((x)^y)^{\sqrt{3}}$ and $((y)^x)^{\sqrt{3}}$ when $x = 3\sqrt{3}$ and $y = \sqrt{3}$.
$((3\sqrt{3})^{\sqrt{3}})^{\sqrt{3}} = (3\sqrt{3})^3 =$
$3^3(\sqrt{3})^3 = 3^3(\sqrt{3})^2(\sqrt{3}) =$
$27(3\sqrt{3}) = 81\sqrt{3};$
$((\sqrt{3})^{3\sqrt{3}})^{\sqrt{3}} = (\sqrt{3})^9 =$
$(\sqrt{3})^8(\sqrt{3}) = ((\sqrt{3})^2)^4(\sqrt{3}) =$
$3^4(\sqrt{3}) = 81\sqrt{3}$

WHAT DO WE KNOW ABOUT π AND e?

Because π and e make so many different appearances in important mathematical situations, people have become fascinated with the numbers and have subjected them to a great deal of study. We know many meaningful facts about the relationships between π and e. We also know many meaningless facts. That is, some information about π and e is simply not very useful.

One example of a meaningless, yet curious fact is that the sixteenth and seventeenth decimal digits of π equal the sixteenth and seventeenth decimal digits of e. Do you know what those digits are? (They are 23.) Another more interesting fact was discovered by R. G. Duggleby, a biochemist at the University of Ottawa. Duggleby discovered that $\pi^4 + \pi^5$ is *almost* equal to e^6. Had the two values been exactly equal, the discovery would have been a remarkable coincidence. As it turns out, the two values are equal correct to four decimal places—a noteworthy coincidence in itself. Can you find the two values correct to five decimal places to verify Duggleby's discovery? You will need a calculator with more than eight digits in its readout. (The sum $\pi^4 + \pi^5$ is 403.42878 correct to five decimal places. The value of e^6 is 403.42879 correct to five decimal places.)

Knowing how two numbers compare can be very helpful information, particularly when the comparison is not immediately obvious. Duggleby discovered that $e^6 > \pi^4 + \pi^5$. A more interesting comparison is the following one.

$$e^\pi > \pi^e$$

The inequality can be verified using a scientific calculator. What are the values of e^π and π^e correct to 6 decimal places? (The value of e^π is about 23.140692 and the value of π^e is about 22.459157.)

WHAT DO WE KNOW ABOUT π and e?   1

π = 3.141592653589793238 . . .

e = 2.718281828459045235 . . .

What are the sixteenth and seventeenth digits of π? _23_

What are the sixteenth and seventeenth digits of e? _23_

Compare $\pi^4 + \pi^5$ to e^6.

    What is $\pi^4 + \pi^5$ correct to 5 decimal places? _403.42878_

    What is e^6 correct to 5 decimal places? _403.42879_

Compare e^π to π^e.

    What is e^π correct to 6 decimal places? _23.140692_

    What is π^e correct to 6 decimal places? _22.459157_

146   A VERY SPECIAL NUMBER CALLED e Copyright © 1983 by Dale Seymour Publications

---

If you complete the table on page 2, you should be able to accept the fact that $x^{1/x}$ reaches its *maximum* value when $x = e$ (a fact proved in calculus). Using this fact, you can easily prove that $e^{\pi} > \pi^{e}$ without using a calculator. The proof is at the bottom of page 2. Can you supply the reasons for each step?

Mathematicians have accumulated a great deal of information about the nature of π and e. You have seen only a few examples from the mass of knowledge we have. Still, there is much we don't know. We know that π and e are both transcendental and irrational, but we don't yet have that information about some of the numbers derived from π and e. It has been proved that e^{π} is an irrational, transcendental number. On the other hand, no one knows whether or not π^{e}, πe, or $\pi + e$ are irrational. Despite the volume of discovered facts about π and e, there are many significant questions that remain unanswered.

WHAT DO WE KNOW ABOUT π and e?		2
x	$\frac{1}{x}$	$x^{\frac{1}{x}}$
1	1	1.0000000
2	0.5	1.4142136
2.1	0.4762	1.4237632
2.5	0.4	1.4426999
2.9	0.3704	1.4436024
2.71	0.3690	1.4446654
2.718	0.3679	1.4446679
2.7182	0.36789	1.4446679
2.71828	0.367880	1.4446679
2.718281	0.3678796	1.4446679
2.7182818	0.3678795	1.4446679
2.8	0.3571429	1.4444393
2.9	0.3448276	1.4436024
3.0	0.3	1.4422496

$x^{\frac{1}{x}}$ reaches its maximum value at $x = e$

$e^{\frac{1}{e}} > \pi^{\frac{1}{\pi}}$ $x^{\frac{1}{x}}$ reaches max. at $x = e$

$\left(e^{\frac{1}{e}}\right)^{e\pi} > \left(\pi^{\frac{1}{\pi}}\right)^{e\pi}$ positive exponentiation preserves inequality

$e^{\pi} > \pi^{e}$ power of a power property

Copyright © 1983 by Dale Seymour Publications   A VERY SPECIAL NUMBER CALLED e 141

EXERCISES

1. Find the decimal value of 355/113 correct to seven decimal places. Describe the relationship between this number and π.

2. Find the decimal value of 553/312 correct to six decimal places.

3. Find the value of $\sqrt{\pi}$ correct to six decimal places.

4. Use the results from problems 2 and 3 to describe the relationship between 553/312 and $\sqrt{\pi}$.

ANSWERS

1. 3.1415929; 355/113 approximates π correct to six decimal places.

2. 1.7724359

3. 1.7724539

4. 553/312 approximates $\sqrt{\pi}$ correct to four unrounded decimal places.

A VERY SPECIAL NUMBER CALLED e 29

EULER'S FORMULA

What do the two most famous transcendental numbers, π and e, have in common with the imaginary unit i, and the additive inverse of the multiplicative identity for the real numbers? How are $\lim_{x \to \infty} (1 + (1/x))^x$, the ratio of the circumference of a circle to its diameter, and a solution to $x^2 + 1 = 0$ related? What is one of the most amazing and useful coincidences in mathematics?

The answer to all these questions is Euler's Formula, discovered by Leonard Euler in the 16th century and considered by many to be the most elegant formula in all of mathematics. Stated in the following way, the formula provides a link between exponential functions and circular functions.

$$e^{ix} = \cos x + i \sin x \qquad \text{where } x \text{ is a real number}$$

By substituting π for x and evaluating terms in the equation, you will obtain an important relationship among e, π, i, and 1.

$$e^{i\pi} = \cos \pi + i \sin \pi$$
$$e^{i\pi} = (-1) + i(0)$$
$$e^{i\pi} = -1$$

There is another, more visual way of obtaining this relationship. It involves graphing a sequence of approximations for $e^{i\pi}$ in the complex plane. One of e's appearances is in the following mathematical statement.

$$e^x = 1 + x + \frac{x^2}{2!} + \frac{x^3}{3!} + \frac{x^4}{4!} + \cdots$$

where x is any complex number

The statement means that, for any given value of x, the value of e^x can be approximated more and more closely by adding more and more terms to the sum given. For example, the sum $1 + x + (x^2/2!)$ gives a rough approximation of e^x.

EULER'S FORMULA 1

$$e^{ix} = \cos x + i \sin x$$
where x is a real number

Use Euler's formula to find the value of $e^{i\pi}$.

$$e^{i\pi} = \cos \pi + i \sin \pi$$
$$= (-1) + i(0)$$
$$= -1$$

Euler's formula establishes the relationship among the four important numbers that are defined below. Which is which?

$\underline{e}$ $\lim\limits_{x \to \infty} \left(1 + \frac{1}{x}\right)^x$

$\underline{\pi}$ $\dfrac{\text{circumference}}{\text{diameter}}$

$\underline{-1}$ additive inverse of the multiplicative identity for real numbers

$\underline{i}$ a solution to $x^2 + 1 = 0$

EULER'S FORMULA 2

Approximate each partial sum. Use 3.14 for π.

Term of Sequence	Value
1	$1 + 0.00i$
$1 + \pi i$	$1 + 3.14i$
$1 + \pi i + \frac{(\pi i)^2}{2!}$	$-3.93 + 3.14i$
$1 + \pi i + \frac{(\pi i)^2}{2!} + \frac{(\pi i)^3}{3!}$	$-3.93 - 2.03i$
$1 + \pi i + \frac{(\pi i)^2}{2!} + \frac{(\pi i)^3}{3!} + \frac{(\pi i)^4}{4!}$	$0.12 - 2.03i$
$1 + \pi i + \frac{(\pi i)^2}{2!} + \frac{(\pi i)^3}{3!} + \frac{(\pi i)^4}{4!} + \frac{(\pi i)^5}{5!}$	$0.12 + 0.52i$
$1 + \pi i + \frac{(\pi i)^2}{2!} + \cdots + \frac{(\pi i)^6}{6!}$	$-1.21 + 0.52i$
$1 + \pi i + \frac{(\pi i)^2}{2!} + \cdots + \frac{(\pi i)^7}{7!}$	$-1.21 - 0.08i$
$1 + \pi i + \frac{(\pi i)^2}{2!} + \cdots + \frac{(\pi i)^8}{8!}$	$-0.98 - 0.08i$
$1 + \pi i + \frac{(\pi i)^2}{2!} + \cdots + \frac{(\pi i)^9}{9!}$	$-0.98 + 0.01i$
$1 + \pi i + \frac{(\pi i)^2}{2!} + \cdots + \frac{(\pi i)^{10}}{10!}$	$-0.95 + 0.01i$

The sum $1 + x + (x^2/2!) + (x^3/3!) + (x^4/4!) + \ldots + (x^{100}/100!)$ gives a much better approximation for e^x.

Suppose x is $i\pi$. The equation becomes

$$e^{i\pi} = 1 + \pi i + \frac{(\pi i)^2}{2!} + \frac{(\pi i)^3}{3!} + \cdots$$

By graphing the sequence of *partial sums* of the series in the equation, you will get a visual interpretation of the approximations for $e^{i\pi}$. (A sequence of partial sums of a series consists of: the first term, the sum of the first two terms, the sum of the first three terms, and so on.) The table on page 2 is laid out to help you determine the first ten partial sums of the sequence.

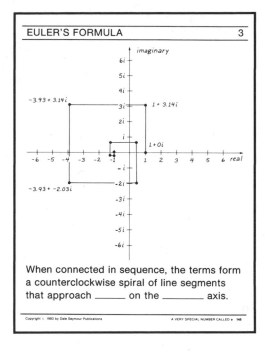

EULER'S FORMULA   3

When connected in sequence, the terms form a counterclockwise spiral of line segments that approach _____ on the _____ axis.

EXERCISES

1. Express each of the following exponential expressions as a complex number in simplest form by using Euler's Formula.

   a. $e^{(\pi i)/2}$
   b. $e^{2\pi i}$
   c. $e^{(\pi/6)(i)}$
   d. $e^{(\pi/3)(i)}$
   e. $e^{(5/6)(\pi i)}$
   f. $e^{(4/3)(\pi i)}$
   g. $e^{(11/6)(\pi i)}$
   h. $e^{(9/4)(\pi i)}$
   i. $e^{3\pi i}$

2. Express each of the following sums as a power of e by using Euler's Formula.

   a. $\cos(2/3)\pi + i \sin(2/3)\pi$
   b. $\cos(\pi/4) + i \sin(\pi/4)$
   c. $\cos(3/2)\pi + i \sin(3/2)\pi$
   d. $\cos 4\pi + i \sin 4\pi$
   e. $(\sqrt{3}/2) + i(1/2)$ (Use the least value of x.)

ANSWERS

1a. $0 + i$
1b. $1 + 0i$
1c. $(\sqrt{3}/2) + (1/2)i$
1d. $(1/2) + (\sqrt{3}/2)i$
1e. $-(\sqrt{3}/2) + (1/2)i$
1f. $-(1/2) - (\sqrt{3}/2)i$
1g. $(\sqrt{3}/2) - (1/2)i$
1h. $(\sqrt{2}/2) + (\sqrt{2}/2)i$
1i. $-1 + 0i$

2a. $e^{(2/3)(\pi i)}$
2b. $e^{(\pi/4)(i)}$
2c. $e^{(3/2)(\pi i)}$
2d. $e^{4\pi i}$
2e. $e^{(\pi/6)(i)}$

PUTTING EULER'S FORMULA TO WORK

Before Euler's time, mathematicians knew that the product of any two pure imaginary numbers is a real number. It was Euler, though, who discovered that a pure imaginary number raised to a pure imaginary power is a real number. Specifically, Euler found that i^i is a real number and, using his formula, found its value.

$$i^i = e^{-\left(\frac{\pi}{2}\right)}$$

Using this relationship, you can compute the decimal value of i^i to any degree of accuracy you choose. Page 1 gives the decimal value of i^i correct to ten decimal places. Using a scientific calculator and 2.7182818 for e, what approximation will you get? (The value is 0.2078796.) There appears to be no pattern to the decimal digits for i^i and, in fact, that is the case. In other words, i^i is irrational.

So far, you have only computed *one* value for i^i. There are more. As a matter of fact, i^i has an infinity of real number values. You can find the values by first taking another look at Euler's Formula and generalizing it. Remember that $e^{ix} = \cos x + i \sin x$ and that both cosine and sine are *periodic* functions with periods of 2π. (A function f is said to be periodic and have period p if $x + p$ is in the domain of the function whenever x is in the domain and $f(x + p) = f(x)$.) Since the periods of cosine and sine are 2π, Euler's Formula tells you that e^{ix} is equal to $e^{ix + 2k\pi}$ where k is any integer. What does this fact tell you about the values of i^i? It tells you this:

$$i^i = e^{-\left(\frac{\pi}{2}\right) + 2k\pi} \qquad \text{where } k \text{ is an integer}$$

The chart on page 1 shows some decimal approximations for values of i^i. Try to find the others yourself. Use 2.71828 for e.

PUTTING EULER'S FUNCTION TO WORK   1

$$i^i = e^{-\left(\frac{\pi}{2}\right)} \approx 0.2078795763\ldots$$

k	i^i or $e^{-\left(\frac{\pi}{2}\right)+2k\pi}$ value	decimal approximation	$\frac{1}{i^i}$ or ____ value	decimal approximation
0	$e^{-\left(\frac{\pi}{2}\right)}$	0.2078798	$e^{\frac{\pi}{2}}$	4.8104774
1	$e^{\frac{3\pi}{2}}$	111.31743	$e^{\frac{5\pi}{2}}$	2575.9703
-1	$e^{-\left(\frac{5\pi}{2}\right)}$	0.0003882	$e^{-\left(\frac{3\pi}{2}\right)}$	0.0089833
2	$e^{\frac{7\pi}{2}}$	59609.732	$e^{\frac{9\pi}{2}}$	1379410.4
-2	$e^{-\left(\frac{9\pi}{2}\right)}$	0.0000007	$e^{-\left(\frac{7\pi}{2}\right)}$	0.0000168
3	$e^{\frac{11\pi}{2}}$	31920512	$e^{\frac{13\pi}{2}}$	7.3866×10^8
-3	$e^{-\left(\frac{13\pi}{2}\right)}$	1.3538×10^{-9}	$e^{-\left(\frac{11\pi}{2}\right)}$	3.1328×10^{-8}

The number $i^{1/i}$ has an infinite number of real number values, too. By using Euler's Formula and the fact that cosine and sine are periodic functions, you can find an exponential expression for the real number values. Try to discover what that expression is and, then, compute the various decimal approximations asked for in the table on page 1.

As you can see, Euler's Formula is a very useful mathematical tool. Not only does it help you determine important relationships between numbers, but it helps you determine the specific character of given numbers and actually helps you calculate decimal approximations of those numbers to any degree of accuracy you might need.

EXERCISES

1. Use Euler's Formula to show that $i^i = e^{-(\pi/2)}$.
   HINT: $e^{-(\pi/2)} = e^{(\pi/2)i^2} = e^{((\pi/2)i)i}$.

2. Use Euler's Formula to show that $i^{1/i} = e^{\pi/2}$.
   HINT: $e^{\pi/2} = e^{(\pi/2)(i/i)} = e^{((\pi/2)i)(1/i)}$.

ANSWERS

1. $e^{-(\pi/2)}$
$$= (e^{(\pi/2)i})^i$$
$$= (\cos(\pi/2) + i\sin(\pi/2))^i$$
$$= (0 + i(1))^i$$
$$= i^i$$

2. $e^{\pi/2}$
$$= (e^{(\pi/2)i})^{1/i}$$
$$= (\cos(\pi/2) + i\sin(\pi/2))^{1/i}$$
$$= (0 + i(1))^{1/i}$$
$$= i^{1/i}$$

APPROXIMATING *e*

Using the formula $f(x) = (1 + (1/x))^x$ you can generate a set of ordered pairs called a *sequence function*. First, let the domain be the set of positive integers. Starting with 1 for x, substitute consecutive positive integers into the expression $(1 + (1/x))^x$ to determine the corresponding elements of the range of the sequence function. The first five pairs of values are given in the table on page 1. Use a calculator to find the remaining range values in the table.

The complete set of ordered pairs obtained in the manner described and listed in order is called a sequence function. The formula $f(x) = (1 + (1/x))^x$ is called a *sequence function generator*.

As you continue to build the sequence function in this lesson, you will make a very interesting discovery. The sequence of values in the range get closer and closer together. The values approach a number that should be very familiar to you by now. That number is *e*. In fact, mathematicians often use this sequence to define *e*. They say,

$$e = \lim_{x \to \infty} \left(1 + \frac{1}{x}\right)^x \qquad \text{where } x \text{ is a positive integer}$$

This statement means that by choosing a sufficiently large value for x, you can get a value as close to *e* as you wish.

Look at the table of values you completed. How great a value of x must you choose to obtain a value for *e* that is correct to 7 decimal digits? (80,000.)

Using today's computers, it is possible to obtain very large values for x. Consequently, it is possible to obtain very exact approximations for *e*.

APPROXIMATING *e*	1
x	$\left(1 + \frac{1}{x}\right)^x$
1	2.0000000
2	2.2500000
3	2.3703704
4	2.4414063
5	2.4883200
8	2.5657845
20	2.6532977
80	2.6991164
100	2.7048138
1000	2.7169239
50,000	2.7182547
70,000	2.7182597
80,000	2.7182818

$$e = \lim_{x \to \infty} \left(1 + \frac{1}{x}\right)^x$$

EXERCISES

1. Use a calculator to evaluate $(1 + (1/x))^x$ for $x = 90$.

2. Evaluate $(1 + (1/x))^x$ for $x = 500$.

3. Evaluate $(1 + (1/x))^x$ for $x = 25{,}000$.

4. Evaluate $(1 + (1/x))^x$ for $x = 60{,}000$.

5. Evaluate $(1 + (1/x))^x$ for $x = 75{,}000$.

6. Evaluate $(1 + (1/x))^x$ for $x = 79{,}000$.

7. How accurate an approximation for e do you obtain if you use 10,000 for x?

ANSWERS

1. 2.7033324

2. 2.7155682

3. 2.7182302

4. 2.718222

5. 2.7182139

6. 2.7182329

7. The approximation is correct to 3 decimal places.

FOUR

ALGEBRAIC FUNCTIONS*

More than anything else, second-year algebra is the study of functions. At least seventeen different kinds of functional relationships are studied—from quadratic and cubic functions to circular and exponential functions. There is, of course, good reason to study functions because so much of the real world exhibits functional relationships. Often we are able to gain clearer insight into a situation by studying its functional relationships.

In second-year algebra, students learn to recognize various functions and relations by identifying characteristics—both algebraic and graphic. The following lessons show some interesting applications for functions students have learned about—applications that show the broad range of circumstances in which functions are useful and meaningful.

*The information in the lessons that follow are based on ideas from:

"The Neutral Theory of Molecular Evolution" by Motoo Kimura in the November 1979 issue of *Scientific American*, page 104.

"World Uranium Resources" by Kenneth S. Deffeyes and Ian D. MacGregor in the January 1980 issue of *Scientific American*, page 74.

"World Coal Production" by Edward D. Griffith and Alan W. Clarke in the January 1979 issue of *Scientific American*, page 46.

"The Coupled Motions of Piano Strings" by Gabriel W. Weinrich in the January 1979 issue of *Scientific American*, page 126.

"Acid Rain" by Gene E. Likens and others in the October 1979 issue of *Scientific American*, page 46.

"A Method of Discovery, II" by W. W. Sawyer in the January 1959 issue of *The Mathematics Student Journal*.

The third edition of *Using Advanced Algebra* by K. J. Travers, L. C. Dalton, and V. F. Brunner, page 308.

"Dots and Cubes" by Bryan L. Eaton in the February 1974 issue of *The Mathematics Teacher*, pages 161–164.

An Introduction to Abstract Mathematical Systems by David M. Burton, pages 21–25.

DETERMINING A FUNCTION

In algebra, one way of determining how pairs of values are related is to graph them to see whether they take the shape of a familiar line or curve. This is an important and helpful technique any time you are interested in quantifying the relationship between variables, and it is a technique used in a variety of real-world situations.

Suppose you are a business manager. You manage people who do part-time phone sales. The sales people are paid a small flat fee each week as long as they make at least 15 phone calls. In addition, they are paid a fixed commission for each item they sell. The table of values on page 1 gives weekly paycheck amounts for various numbers of items sold. Plot the points on a graph. The points should suggest a straight-line graph that slopes upward from the bottom left to the upper right. Draw a line through the points you've graphed. This line represents the relationship between weekly pay in dollars and number of items sold. By choosing several points on the line you can determine its equation.

The general expression for a straight line is $y = mx + b$ where m is the slope and b is the y-intercept. First use your points to determine the slope by finding the change in y-values with respect to the change in x-values. Once you've found slope, you can substitute that value and the coordinates of one point into your equation to determine the value of b. What is the equation?

With the particular data you were given, you should have obtained $y = 8x + 5$. This equation is a meaningful and descriptive equation for the sales situation under discussion. In this case, 5 represents the flat weekly fee each sales representative receives and 8 represents the commission for each item sold.

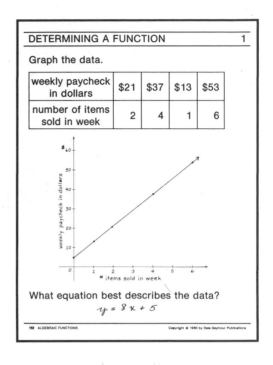

DETERMINING A FUNCTION   1

Graph the data.

weekly paycheck in dollars	$21	$37	$13	$53
number of items sold in week	2	4	1	6

What equation best describes the data?
$y = 8x + 5$

162 ALGEBRAIC FUNCTIONS   Copyright © 1988 by Dale Seymour Publications

This first example on phone sales was presented merely to remind you how to find an equation from graphed data and to point out a simple situation that could be described by a linear equation. Most business people would understand the relationship in the example and would not need to graph the data to determine the equation.

There are many situations in which it is truly helpful to graph data in order to show the quantifiable relationship between variables. A medical researcher may be reasonably sure that height is related to age for a given population, but may not know the most nearly exact equation to use for describing that particular population. Or a scientist may not know whether or not changes in the amino acids of molecules are uniform over long periods of time, but may have a large amount of data to test such a hypothesis. Only rarely do the data in these real-life examples take shape as nicely as data in an algebra text usually does, but often the graphs are quite suggestive, giving researchers important clues for their work.

Look at the data graphed on page 2. It does not exactly follow a straight line, but it tends to. Draw a line through the points that it seems to follow. Then find an equation for your line. Do you see how your equation could be used to make future predictions about the same experimental situation?

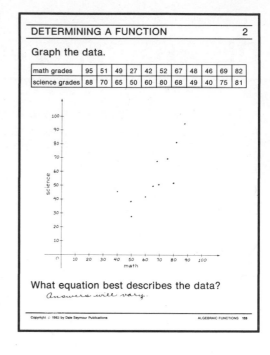

DETERMINING A FUNCTION   2

Graph the data.

| math grades | 95 | 51 | 49 | 27 | 42 | 52 | 67 | 48 | 46 | 69 | 82 |
| science grades | 88 | 70 | 65 | 50 | 60 | 80 | 68 | 49 | 40 | 75 | 81 |

What equation best describes the data?
Answers will vary.

EXERCISES

1. Graph the following set of data. Determine a linear equation that approximates the data.

x	33	45	46	20	40	30	38	22	52	44	55
y	4	7	8	1	6	3	5	2	9	6	10

2. Graph the following set of data. Determine a linear equation that approximates the data.

x	1	3	5	8	9	10	12	15
y	1.0	2.0	3.0	4.5	5.0	5.5	6.5	8.0

ANSWERS

1. $y = 4.3x + 14.5$

2. $y = 0.5x + 0.5$

ALGEBRAIC GRAPHS

Graphs are convenient tools for displaying data and showing relationships in a concise visual form. By making a quick study of back issues of *Scientific American,* I found a variety of graphs that show algebraic relationships you commonly study in high school algebra. They show a parabola, addition of functions, periodic functions, and logarithmic functions. Page 1 asks you to look for the four different graphs. Notice the variety of applications for these graphs.

The general equation for a parabola of the type shown on page 1 is $y = a(x - h)^2 + k$. In this graph, the parabola opens downward. What does that tell you about the value of a? (It is negative.)

The coal data shows addition of functions. To see how the graph is an application of this concept, choose any vertical line and add the ordinates; that is, add the amount of coal mined from surface mines to the amount mined from underground mines in a given year. The sum gives a point on the graph that shows the total amount of bituminous coal mined in the United States that year.

Do you see how the graph of strings displacements also shows addition of functions? In this case, the functions are periodic too.

The last graph shows an application of logarithms. What is a negative logarithm, you ask? Well, look at the graphs on page 2. The equation for f is $f(x) = \log_{10} x$ where $0 < x \leq 1$. The equation for g is $g(x) = -\log_{10} x$ where $0 < x \leq 1$. The graph of g is the mirror image of f through the x-axis. And for every value of x in the domain, $g(x) = -f(x)$.

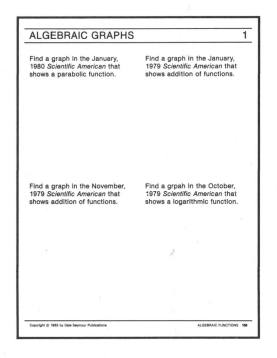

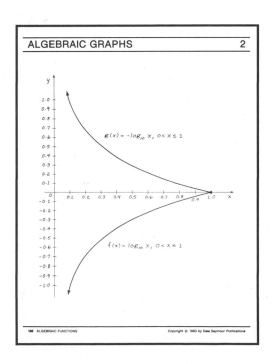

EXERCISES

1. For each parabola whose equation is given, find the maximum value, the maximum point, and name the axis of symmetry.

    a. $y = -2(x - 3)^2 + 4$
    b. $y = -(x + 5)^2 + 3$
    c. $y = -(5/4)(x - 7)^2 + (2/3)$
    d. $y = -(1/2)(x + 4)^2 - 3$

2. How would you describe the graph of the quadratic function whose equation is $y = 2(x - 4)^2 + 5$?

3. Graph the following functions on the same coordinate axes. Give the equation for the graph of **c**.

    a. $f(x) = (1/2)x + 1$
    b. $g(x) = (x - 2)^2$
    c. $(f + g)(x)$

4. Graph the following functions on the same coordinate axes. Explain how they are related.

    a. $f(x) = (1/3)x + 2$
    b. $-f(x)$

5. Graph the following functions on the same coordinate axes. Explain how they are related.

    a. $g(x) = (1/2)(x - 2)^2 + 1$
    b. $-g(x)$

ANSWERS

1a. 4, (3, 4), $x = 3$
  b. 3, $(-5, 3)$, $x = -5$
  c. 2/3, (7, 2/3), $x = 7$
  d. -3, $(-4, -3)$, $x = -4$

2. The graph is a parabola that opens upward. The minimum value is 5 at $x = 4$.

3. $(f + g)(x) = $ $x^2 - (3\text{-}1/2)x + 5.$

4. The two graphs are reflections of one another through the x-axis. They intersect at $x = -6$.

5. The two graphs are reflections of one another through the x-axis. They are both parabolas.

A QUADRATIC PIZZA FUNCTION

It has probably never occurred to you that cutting a pizza into pieces is mathematical, but it is. Here is a mathematical problem about pizza cutting.

> If you make exactly five cuts in a pizza, what is the *greatest* number of pieces you can obtain?

If your answer is ten, you haven't thought hard enough. Look at the pictures on page 1. They show the maximum number of pieces you can obtain for 0, 1, 2, 3, and 4 cuts. By completing the table, you'll get a good idea for the answer to your problem. (Zero cuts give 1 piece; 1 cut gives 2 pieces; 2 cuts make 4 pieces; 3 cuts can give 7 pieces; and 4 cuts can give 11 pieces.)

Look at the sequence of numbers representing the maximum number of pieces for various cuts. The first two numbers differ by 1. By how much do the second and third numbers differ? Write your answer on page 1. Find the differences for the other numbers in your sequence. (The first row of differences is 1, 2, 3, 4.) By how much do the numbers in your sequence of differences differ? (They each differ by 1.) Use all the information you've obtained to answer the original pizza problem. What is the maximum number of pieces you can make with five cuts? (The answer is 16.)

Suppose *r* stands for the maximum number of pizza pieces you can get by cutting the pizza *n* times. Then *n* and *r* have the following functional relationship.

$$r = \frac{1}{2}n^2 + \frac{1}{2}n + 1$$

Check your numbers against this formula. It is possible to obtain this quadratic pizza formula from the difference data you found. The secret lies in determining how the first numbers from each row are related to the coefficients 1/2, 1/2, and 1.

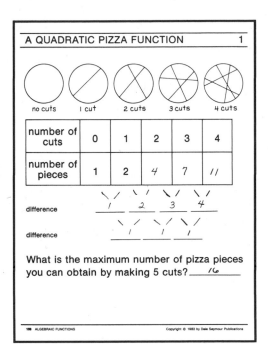

A QUADRATIC PIZZA FUNCTION   1

no cuts   l cut   2 cuts   3 cuts   4 cuts

number of cuts	0	1	2	3	4
number of pieces	1	2	4	7	11

difference   1   2   3   4

difference   1   1   1

What is the maximum number of pizza pieces you can obtain by making 5 cuts? ___16___

ALGEBRAIC FUNCTIONS   Copyright © 1983 by Dale Seymour Publications

Consider a general quadratic function $f(x) = ax^2 + bx + c$. Complete the table on page 2 to find $f(0)$, $f(1)$, $f(2)$, $f(3)$, and $f(4)$. (The value of $f(0)$ is c; $f(1) = a + b + c$; $f(2) = 4a + 2b + c$; $f(3) = 9a + 3b + c$; $f(4) = 16a + 4b + c$.) Then find the two rows of differences. (The differences in the first row are $a + b$, $3a + b$, $5a + b$, and $7a + b$. The differences in the second row are all $2a$.)

Do you see how to use the first terms in each row to find the quadratic relationship? If the first numbers in each row are 1, 1, and 1, respectively, what is the quadratic relationship? (The relationship is $f(x) = (1/2)x^2 + (1/2)x + 1$.) If the first numbers in each row are -2, 3, and 4, respectively, what is the quadratic relationship? (The relationship is $f(x) = 2x^2 + 2x - 2$.)

A QUADRATIC PIZZA FUNCTION   2

$$f(x) = ax^2 + bx + c$$

x	0	1	2	3	4
y	c	$a+b+c$	$4a+2b+c$	$9a+3b+c$	$16a+4b+c$

difference $a+b$ $3a+b$ $5a+b$ $7a+b$

difference $2a$ $2a$ $2a$

If the first numbers in each row are 1, 1, and 1, respectively, what is the quadratic relationship? $f(x) = \frac{1}{2}x^2 + \frac{1}{2}x + 1$

If the first numbers in each row are -2, 3, and 4, respectively, what is the quadratic relationship? $f(x) = 2x^2 + x - 2$

Copyright © 1983 by Dale Seymour Publications   ALGEBRAIC FUNCTIONS   199

EXERCISES

Ask a friend to *think* of a quadratic equation in the form $y = ax^2 + bx + c$. (Your friend should decide on particular values for a, b, and c.) Then ask your friend to tell you the values for y when x is 0, 1, 2, 3, and 4. Your friend should give you the values in order. Write the values on a piece of paper or a chalkboard for your friend (and audience) to see. Find the differences as you did in this lesson. From the differences, determine the original quadratic equation. Any friend who doesn't know the trick will be amazed.

1. What is the quadratic equation if your friend's values are -4, 1, 12, 29, and 52?

2. What is the quadratic equation if your friend's values are -3, 4, 15, 30, and 49?

3. What is the quadratic equation if your friend's values are 8, 7, 16, 35, and 64?

ANSWERS

1. $y = 3x^2 + 2x - 4$

2. $y = 2x^2 + 5x - 3$

3. $y = 5x^2 - 6x + 8$

44   ALGEBRA IN THE REAL WORLD

FINDING CUBICS

For a cubic function $f(x) = ax^3 + bx^2 + cx + d$, if you are given values for $x = 0$, 1, 2, and 3, you can find the numerical coefficients. You use differences, just as you did with quadratic functions. To get an idea of how to do it, first complete the table on page 1 for the general cubic function. Then find the differences. Notice that for cubics you must find *three* sets of differences. Now check your work. The values of the function are c, $a + b + c + d$, $8a + 4b + 2c + d$, $27a + 9b + 3c + d$, and $64a + 16b + 4c + d$. The first row of differences is $a + b + c$, $7a + 3b + c$, $19a + 5b + c$, $37a + 7b + c$. The second row of differences is $6a + 2b$, $12a + 2b$, and $18a + 2b$. The last row of differences is $6a$ and $6a$.

Do you see how to use the differences to find the numerical coefficients for a cubic equation? Try out your ideas for the function in the table at the bottom of the page. (You should get $f(x) = 2x^3 - 3x^2 + 5x + 13$.)

Cubic equations can be used to describe geometric situations. Consider, for example, a cubic structure that is further subdivided into layers of cubes. Page 2 shows three examples. The first example is simply a cubic structure made up of 12 individual segments. The second structure has two layers of eight cubes each and is made from 54 individual segments. Do you see them? How many individual segments make up the third structure, a three-cube by three-cube by three-cube structure? (It uses 144 individual segments.) Try to determine how many individual segments make up structures that have four cubes on a side and five cubes on a side to complete the table on page 2. (The four-cube structure has 300 individual segments and the five-cube structure has 540 individual segments.)

There is an equation that allows you to count the number of individual segments s within a structure as long as you know how many segments are on a side x. As you might guess, the equation is a cubic equation. You can use the difference technique you've learned with the information in the

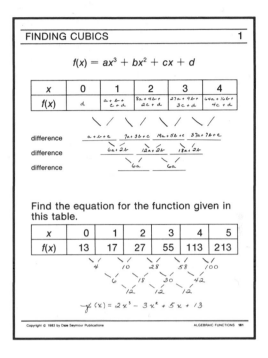

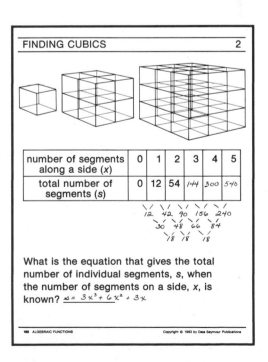

table you completed to find the equation. What is the equation? (It is $s = 3x^3 + 6x^2 + 3x$.)

This particular example of using the difference technique to find an equation should point out how helpful the technique can be. Counting segments from the diagrams is not an easy task and discerning a pattern by a cursory examination of the numbers is very difficult. The difference technique gives a good workable method for finding patterns.

EXERCISES

ANSWERS

1. Determine the cubic equation that could generate the following data.

x	0	1	2	3	4	5
y	-10	-22	-24	-10	26	90

1. $y = x^3 + 2x^2 - 15x - 10$

2. Determine the cubic equation that could generate the following data.

x	0	1	2	3	4	5
y	14	3	-6	5	54	159

2. $y = 3x^3 - 8x^2 - 6x + 14$

3. Find the number of individual segments within a cubic structure that has six segments on a side.

3. 882

4. Find the number of individual segments within a cubic structure that has seven segments on a side.

4. 1344

5. Find the number of individual segments within a cubic structure that has eight segments on a side.

5. 1944

6. Find the number of individual segments within a cubic structure that has nine segments on a side.

6. 2700

A SPECIAL GROUP OF FUNCTIONS

In your study of mathematics, you have learned about a variety of operations that can be used to combine elements of a set. For example, you have learned to combine integers using addition; you have learned to combine elements of a set by intersection of sets; you have learned to combine functions by using composition of functions. In short, you have learned about algebraic systems—sets of objects along with some operations for combining them.

A *group* is a very simple algebraic system. A group consists of a set of elements, integers for example, along with a rule of combination defined for pairs of elements (called a single binary operation), say addition, for which the following properties are satisfied.

1. Closure: The combination of any two elements of the set also belongs to the set.

2. Associativity: When combining elements of the set, they can be grouped in any way. $(a * (b * c) = (a * b) * c)$

3. Identity Element: There is an element of the set that when combined with any other element of the set leaves that element unchanged.

4. Inverses: For any element of the set, there is another element that combines with it to produce the identity.

The integers under addition form a group. What is the identity element? (It is 1.) How do you find the additive inverse of an integer? For example, what are the inverses of 5 and −7? (The additive inverse of an integer is the opposite of the number; thus, the additive inverse of 5 is −5 and the additive inverse of −7 is −(−7) or 7.)

A SPECIAL GROUP OF FUNCTIONS   1

A group consists of a set of elements and a rule of combination, *,
(binary operation) defined for pairs of elements for which the
following properties are satisfied.

- **CLOSURE**
  The combination of any two elements of
  the set also belongs to the set.

- **ASSOCIATIVITY**
  When combining elements of the set, they
  can be grouped in any way.
  e.g. $a * (b * c) = (a * b) * c$

- **IDENTITY ELEMENT**
  There is an element of the set that when
  combined with any other element of the set
  leaves that element unchanged.

- **INVERSES**
  For any element of the set, there is another
  element that combines with it to produce
  the identity.

The integers under addition form a group.
What is the identity element?   *1*

How do you find the inverse of an element in that group?
*take the opposite (negative)
of the number*

164 ALGEBRAIC FUNCTIONS   Copyright © 1983 by Dale Seymour Publications

The elements of a group are not necessarily numbers. One interesting group is made up of six different functions that can be combined using composition of functions. The real-valued functions are defined by the equations $f_1(x) = x$, $f_2(x) = 1/x$, $f_3(x) = 1 - x$, $f_4(x) = (x - 1)/x$, $f_5(x) = x/(x - 1)$, and $f_6(x) = 1/(1 - x)$.

On page 2, there is an operations table for the group. When completed, it shows the results of combining pairs of elements in the group. For example, if you combine f_6 and f_4 you get f_1 because $(f_6 \cdot f_4)(x) = f_6(f_4(x)) = f_6((x - 1)/x) = 1/[1 - (x - 1)/x] = x = f_1(x)$. Try to complete the operations table. Be careful about order. Notice that $f_3 \cdot f_2 = f_4$, but $f_2 \cdot f_3 = f_6$! Once you've completed the table, decide what the identity element for this special group of functions is. Because this set of functions forms a group, every element has an inverse. Can you find the inverse of f_5? (The identity element is f_1. The inverse of f_5 is f_5. In fact, every element is its own inverse except for f_4 and f_6 which are inverses of each other.)

Some groups have *subgroups*. A subset of a group is said to be a subgroup if, under the group's operation, the subset itself forms a group. The group of functions you just investigated has a subgroup consisting of f_1, f_4, and f_6. Complete the operations table for this subgroup and use your results to determine the identity element and the inverses for each function. (As with the complete group, the identity element is f_1 and is its own inverse and f_4 and f_6 are inverses of each other. The correct operations can be determined from the original group table.)

A SPECIAL GROUP OF FUNCTIONS   2

The set of functions $\{f_1, f_2, f_3, f_4, f_5, f_6\}$ forms a group under composition of functions where the functions are defined as follows.

$$f_1(x) = x \qquad f_4(x) = \frac{x-1}{x}$$

$$f_2(x) = \frac{1}{x} \qquad f_5(x) = \frac{x}{x-1}$$

$$f_3(x) = 1 - x \qquad f_6(x) = \frac{1}{1-x}$$

Complete the operations table for this group.

*	f_1	f_2	f_3	f_4	f_5	f_6
f_1	f_1	f_2	f_3	f_4	f_5	f_6
f_2	f_2	f_1	f_6	f_5	f_4	f_3
f_3	f_3	f_4	f_1	f_2	f_6	f_5
f_4	f_4	f_3	f_5	f_6	f_2	f_1
f_5	f_5	f_6	f_4	f_3	f_1	f_2
f_6	f_6	f_5	f_2	f_1	f_3	f_4

What is the identity element? ___f_1___

What is the inverse of f_5? ___f_5___

A SPECIAL GROUP OF FUNCTIONS   3

The set of functions $\{f_1, f_4, f_6\}$ forms a subgroup.

Complete the operations table.

*	f_1	f_4	f_6
f_1	f_1	f_4	f_6
f_4	f_4	f_6	f_1
f_6	f_6	f_1	f_4

What is the identity element? ___f_1___

Name the inverse for each function.

inverse of f_1 ___f_1___

inverse of f_4 ___f_6___

inverse of f_6 ___f_4___

The graphs of the functions f_1, f_4, and f_6 in the coordinate plane will give you more insight into the nature of this subgroup. The graphs of f_4 and f_6 are hyperbolas. How are they related? (They're mirror images about the line formed by the identity function f_1. Thus, f_4 and f_6 are inverse functions of one another.)

I think this is a neat little subgroup because, with only three elements, it illustrates not only group properties but inverse relation, inverse function, 1-1 function, identity function, rational function, hyperbola, and linear function. Quite a unique application indeed!

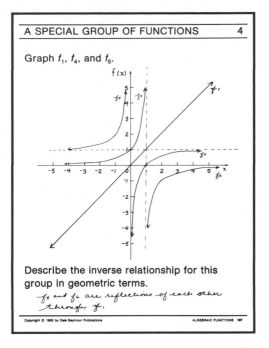

A SPECIAL GROUP OF FUNCTIONS　　4

Graph f_1, f_4, and f_6.

Describe the inverse relationship for this group in geometric terms.

f_4 and f_6 are reflections of each other through f_1.

EXERCISES

1. Which of the following algebraic systems are groups?

   a. whole numbers under addition
   b. rational numbers under addition
   c. integers under multiplication
   d. nonzero rational numbers under multiplication

2. Show that $f_4 \cdot (f_6 \cdot f_4) = (f_4 \cdot f_6) \cdot f_4$.

3. Does the group of functions $f_1, f_2, f_3, f_4, f_5, f_6$ have any subgroups other than the one mentioned in this lesson? If so, name one.

ANSWERS

1. b and d

2. $f_4 \cdot (f_6 \cdot f_4) = f_4 \cdot (f_1) = f_4$;
   $(f_4 \cdot f_6) \cdot f_4 = (f_1) \cdot f_4 = f_4$

3. Yes, there are other subgroups. The sets are $\{f_1\}$, $\{f_1, f_2\}$, $\{f_1, f_3\}$, and $\{f_1, f_5\}$.

FIVE

THE GOLDEN RATIO*

In mathematics there are certain numbers that appear over and over again—often quite unexpectedly. Two such numbers are e and pi. Another number is $(1 + \sqrt{5})/2$, called the Golden Section or Golden Ratio. The variety of its appearances, especially within natural phenomena, have made the Golden Ratio a topic of intense interest ever since the ancient Greeks. It has been studied in both algebraic and geometric forms. It has been observed in contexts as abstract as continued fractions and as mundane as apples.

The next few lessons present several of the Golden Ratio's significant applications, but they can only hint at the mystery and appeal of this popular number.

*The information in the lessons that follow are based on ideas from:

Introduction to Geometry by H. S. M. Coxeter, pages 160–161 and 166.

The Divine Proportion: A Study in Mathematical Beauty by H. E. Huntley, pages 24–25, 28–30, 61, 140, and 170–171.

Brian Yanny's solution to the special triangle which appeared in the September 1980 issue of *The Point Subset* published by the Department of Mathematics of the University of Wisconsin, Stevens Point.

The Golden Section and Related Curiosa by Garth E. Runion, pages 21–25, 35–36, 52–53, 64–67.

The third edition of *Using Advanced Algebra* by K. J. Travers, L. C. Dalton, and V. F. Brunner.

WHAT IS
THE GOLDEN RATIO?

MATHEMATICS CONTENT:
Proportions,
Quadratic Equations,
Irrational Numbers

Suppose a line segment $\overline{RT}$ contains a point S so that the following proportion holds.

$$\frac{RT}{RS} = \frac{RS}{ST}$$

(We say that S divides $\overline{RT}$ into extreme and mean ratio or RS is the geometric mean of RT and ST.) The figure on page 1 shows one such example. In this particular case, $\overline{ST}$ is 1 unit long and $\overline{RS}$ is x units long. How long is $\overline{RT}$? (It is $(x + 1)$ units long.)

With the help of a little algebra, you can find a value for x. Substitute values for the lengths of the line segments into the proportion and cross multiply. You will obtain a simple quadratic equation that can be solved using the quadratic formula. There are two roots to the equation; one is positive and one is negative. Only the positive root is a solution to the problem. (Lengths are not negative in value.) The value you obtain, the positive root, is called the *Golden Ratio* or *Golden Section*. (You should have used the quadratic equation $x^2 - x - 1 = 0$ and obtained the value $(1 + \sqrt{5})/2$ for x.)

A *Golden Rectangle* can be constructed in the following way. Draw an *l* by *l* square. Add an additional length *w* to opposite sides so that

$$\frac{l + w}{l} = \frac{l}{w}$$

Then complete the last side.

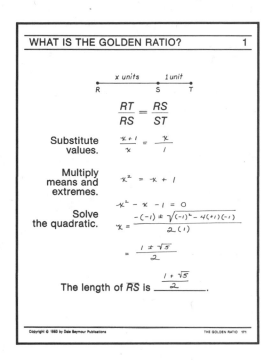

WHAT IS THE GOLDEN RATIO?   1

x units   1 unit

R   S   T

$$\frac{RT}{RS} = \frac{RS}{ST}$$

Substitute
values. $\frac{x+1}{x} = \frac{x}{1}$

Multiply
means and $x^2 = x + 1$
extremes.

$x^2 - x - 1 = 0$

Solve $x = \dfrac{-(-1) \pm \sqrt{(-1)^2 - 4(+1)(-1)}}{2(1)}$
the quadratic.

$= \dfrac{1 \pm \sqrt{5}}{2}$

The length of RS is $\underline{\frac{1 + \sqrt{5}}{2}}$.

The figure on page 2 shows such a rectangle. Try solving the given proportion for l. You should obtain a solution that looks very familiar. (In fact, if you substitute 1 for w and x for l, you will obtain exactly the same solution as you did on page 1.) The ratio l/w is the Golden Ratio. And what is the value of $(l + w)/l$? (It's the Golden Ratio again.) Both the large rectangle ($l + w$ by l) and the small rectangle (l by w) are Golden Rectangles. For both rectangles, the ratio of the length to the width is the Golden Ratio.

Many people say that the proportions of the Golden Rectangle are most pleasing to the eye. In fact, people often unconsciously favor the Golden Rectangle shape over other rectangular shapes. The shape has been used frequently in both art and architecture. The Parthenon in Athens, Greece and the United Nations building in New York are prime examples of buildings whose constructions freely use Golden Rectangles.

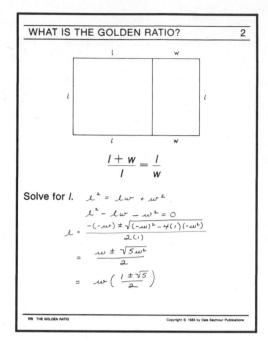

WHAT IS THE GOLDEN RATIO?   2

$$\frac{l + w}{l} = \frac{l}{w}$$

Solve for l.

$$l^2 = lw + w^2$$
$$l^2 - lw - w^2 = 0$$
$$l = \frac{-(-w) \pm \sqrt{(-w)^2 - 4(1)(-w^2)}}{2(1)}$$
$$= \frac{w \pm \sqrt{5w^2}}{2}$$
$$= w\left(\frac{1 \pm \sqrt{5}}{2}\right)$$

172 THE GOLDEN RATIO   Copyright © 1983 by Dale Seymour Publications

EXERCISES

The equation $x^2 - x - 1 = 0$ is called the Golden Quadratic Equation. We use the symbols ϕ and ϕ' to represent the roots of the equation.

$$\phi = \frac{1 + \sqrt{5}}{2} \quad \text{and} \quad \phi' = \frac{1 - \sqrt{5}}{2}$$

1. Find $\phi + \phi'$.

2. Find $\phi \cdot \phi'$.

3. Show that $\phi - 1 = 1/\phi$.

ANSWERS

1. 1

2. -1

3. $\phi - 1 = ((1 + \sqrt{5})/2) - 1$
$= ((1 + \sqrt{5})/2) - (2/2)$
$= (-1 + \sqrt{5})/2;$
$1/\phi = 1/((1 + \sqrt{5})/2)$
$= 2/(1 + \sqrt{5})$
$= (2/(1 + \sqrt{5})) \times$
$\quad ((1 - \sqrt{5})/(1 - \sqrt{5}))$
$= (2(1 - \sqrt{5}))/(-4)$
$= (1 - \sqrt{5})/(-2)$
$= (-1 + \sqrt{5})/2;$
By the transitive property of equality, $\phi - 1 = 1/\phi$.

4. Find a decimal approximation for the Golden Ratio correct to 6 decimal places.

5. Suppose a point C divides a line segment $\overline{AB}$ into two segments, $\overline{AC}$ and $\overline{CB}$, so that $(AC/CB) = (AB/AC)$. Show that each ratio in the proportion is a Golden Ratio.

4. 1.618034

5. $AC/CB = AB/AC$
$= (AC + CB)/AC$
$= 1 + (CB/AC)$
$= 1 + (11/(AC/CB))$
Let $x = AC/CB$. Then the equation becomes $x = 1 + (1/x)$, or $x^2 - x - 1 = 0$. One of the roots of this equation is the Golden Ratio. Thus, by transitivity, both ratios equal the Golden Ratio.

CONTINUED FRACTIONS

A number plus a fraction whose denominator is a number plus a fraction whose denominator is, in turn, a number plus a fraction, and so on is called a *continued fraction*. For example, the following complex fraction is a continued fraction.

$$1 + \cfrac{1}{1 + \cfrac{1}{2 + \cfrac{1}{3}}}$$

(In simplest form, this fraction is 17/10 or 1-7/10.)

The simplest of all *infinite* continued fractions is:

$$1 + \cfrac{1}{1 + \cfrac{1}{1 + \cfrac{1}{1 + \cfrac{1}{1 + \dots}}}}$$

The dots indicate that the successive fractions are continued without end. The dots mean "and so on."

To find the value of such a fraction, it is helpful to look at a sequence of fractions (called *convergents*) obtained by stopping at consecutive stages. The sequence of convergents for our fraction is:

$$1, \ 1 + \frac{1}{1}, \ 1 + \cfrac{1}{1 + \cfrac{1}{1}}, \ 1 + \cfrac{1}{1 + \cfrac{1}{1 + \cfrac{1}{1}}}, \dots$$

The farther out in the sequence you look, the closer an approximation you will get for the infinite continued fraction you're interested in.

The chart on page 1 shows the first few terms of the sequence of convergents and their decimal values. If you study

<table>
<tr><td colspan="2">CONTINUED FRACTIONS   1</td></tr>
</table>

$$1 + \cfrac{1}{1 + \cfrac{1}{1 + \cfrac{1}{1 + \cfrac{1}{1 + \dots}}}}$$

Term	Decimal Value
1st: 1	1
2nd: $1 + \frac{1}{1}$	2
3rd: $1 + \cfrac{1}{1 + \frac{1}{1}}$	1.50
4th: $1 + \cfrac{1}{1 + \cfrac{1}{1 + \frac{1}{1}}}$	$1.\overline{6}$
5th:	1.60
6th:	1.6250
7th:	$1.\overline{615384}$

174 THE GOLDEN RATIO   Copyright © 1983 by Dale Seymour Publications

the sequence of decimal values closely, you will discover that they approach a particular value. That value is the value of the infinite continued fraction.

To find decimal equivalents for continued fractions, you start from the bottom and work up. For example, take the fourth term of the sequence.

$$1 + \cfrac{1}{1 + \cfrac{1}{1 + \cfrac{1}{1}}}$$

The "last" denominator is $1 + (1/1)$ or 2, so you can rewrite the entire fraction as follows.

$$1 + \cfrac{1}{1 + \cfrac{1}{2}}$$

Next, rewrite $1 + (1/2)$ as $(3/2)$. That gives the following expression.

$$1 + \cfrac{1}{\cfrac{3}{2}}$$

Take the reciprocal of $(3/2)$, 1 over $(3/2)$, to get $(2/3)$. Then add it to 1. The value is 1-2/3 or $1.\overline{6}$.

Using this procedure, find the decimal values for the remaining terms in the chart. (The completed sequence is 1, 2, 1.50, $1.\overline{6}$, 1.60, 1.6250, $1.\overline{615384}$, ...) Can you guess what value the terms of the sequence approach? It's the Golden Ratio—1.618034 (correct to 6 decimal places). There's a simple way to convince yourself.

Let x stand for the infinite continued fraction. Then, the entire denominator of the right-hand term also equals x. It's marked with a box on page 2. As a result, you can write a new equation.

$$x = 1 + \frac{1}{x}$$

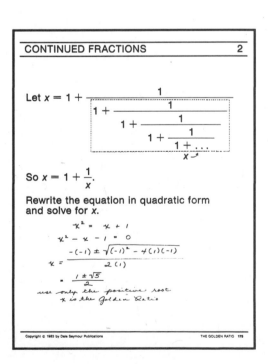

CONTINUED FRACTIONS   2

Let $x = 1 + \cfrac{1}{1 + \cfrac{1}{1 + \cfrac{1}{1 + \cfrac{1}{1 + \cdots}}}}$ x

So $x = 1 + \dfrac{1}{x}$.

Rewrite the equation in quadratic form and solve for x.

$$x^2 = x + 1$$
$$x^2 - x - 1 = 0$$
$$x = \frac{-(-1) \pm \sqrt{(-1)^2 - 4(1)(-1)}}{2(1)}$$
$$= \frac{1 \pm \sqrt{5}}{2}$$

use only the positive root
x is the Golden Ratio

This equation can be written as a quadratic equation (that you have seen before). Simply multiply both sides by x. The quadratic equation is called the Golden Quadratic Equation and its positive root is the Golden Ratio. Can you find the equation and its roots? (The equation is $x^2 = x + 1$ or $x^2 - x - 1 = 0$ and the roots are $(1 \pm \sqrt{5})/2$.)

EXERCISES

ANSWERS

1. Find the rational number in fraction form that equals the following fraction

$$1 + \cfrac{1}{2 + \cfrac{1}{3 + \cfrac{1}{4 + \cfrac{1}{5}}}}$$

1. 225/157

2. Find the rational number in fraction form that equals the following fraction.

$$2 + \cfrac{1}{2 + \cfrac{1}{2 + \cfrac{1}{2 + \cfrac{1}{2}}}}$$

2. 70/29

THE GOLDEN TRIANGLE

Suppose you have an isosceles triangle $\triangle ABC$ with angle measures of 36°, 72°, and 72° and with the equal sides having lengths of 1 unit. How can you determine the length of side $\overline{CB}$—the side opposite the 36° angle?

The lengths of two sides and the measure of the included angle are given. The length of the side opposite the included angle can be found by using the Law of Cosines. Complete the work on page 1 to find an approximate value for the length.

Your work should look something like this series of equations.

$$BC^2 = AC^2 + AB^2 - 2(AB)(AC)\text{Cos } A$$
$$x^2 = 1^2 + 1^2 - 2(1)(1)\text{Cos } 36°$$
$$= 2 - 2 \text{ Cos } 36°$$
$$\approx 2 - 1.618034$$
$$\approx 0.381966$$
$$x \approx \sqrt{0.381966}$$
$$\approx 0.618034$$

A rational approximation of $1/((1 + \sqrt{5})/2)$, the reciprocal of the Golden Ratio, is 0.618034 (correct to 6 decimal places).

What is the value of AC/CB? (The Golden Ratio.) Do you see why this particular isosceles triangle, $\triangle ABC$, is known as the Golden Triangle?

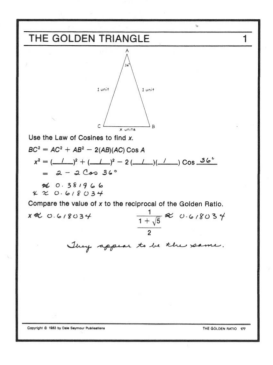

THE GOLDEN TRIANGLE   1

Use the Law of Cosines to find x.

$BC^2 = AC^2 + AB^2 - 2(AB)(AC) \text{ Cos } A$

$x^2 = (\underline{})^2 + (\underline{})^2 - 2(\underline{})(\underline{}) \text{ Cos } \underline{36°}$

$= 2 - 2 \text{ Cos } 36°$

≈ 0.381966

$x \approx 0.618034$

Compare the value of x to the reciprocal of the Golden Ratio.

$x \approx 0.618034$ $\dfrac{1}{\dfrac{1 + \sqrt{5}}{2}} \approx 0.618034$

They appear to be the same.

EXERCISES

1. Find the length of the side opposite the 36° angle in an isosceles triangle for which the two sides including this angle have lengths of 10 units. First predict the answer, then find it.

2. Explain why all Golden Triangles are similar.

3. Prove that AC/CB is the Golden Ratio for $\triangle ABC$ of this lesson.

ANSWERS

1. The answer is ten times the value found for the triangle in the lesson. (The triangle in the lesson is similar to the triangle in this problem, so corresponding sides are proportional. Use the Law of Cosines to verify the answer.)

$$x^2 = 10^2 + 10^2 - 2(10)(10)\text{Cos } 36°$$
$$= 200 - 200 \text{ Cos } 36°$$
$$= 100(2 - 2 \text{ Cos } 36°)$$
$$\approx 100(2 - 1.618034)$$
$$x \approx 6.18034$$

2. Two triangles are similar if their corresponding angles are congruent. Since by definition all Golden Triangles have angle measures of 36°, 72°, and 72°, all Golden Triangles are similar.

3. In $\triangle ABC$, let $\angle C$ and $\angle B$ measure 72°. Draw $\overline{CD}$ with D on $\overline{AB}$ so that $\overline{CD}$ is the bisector of $\angle ACB$. Then $\triangle ACB$ is similar to $\triangle CBD$. Thus, $AC/CB = CB/BD = CB/(AB - AD)$ since corresponding sides of similar triangles are proportional and by the substitution property of equality. By multiplying the means and extremes of the proportion $AC/CB = CB/(AB - AD)$, dividing both sides by $(CB)^2$, and simplifying, you can obtain the following expression. (The reasons for the previous steps are means-extremes product theorem, along with properties of equality, exponents, properties of exponents and fractions.)

$$\left(\frac{AC}{CB}\right)^2 - \left(\frac{AC}{CB}\right)^1 - 1 = 0$$

Using the Quadratic Formula, you will obtain a positive root of $(1 + \sqrt{5})/2$, the Golden Ratio.

THE REGULAR PENTAGON

The regular pentagon bears a very close relationship to the Golden Ratio. In fact, many constructions of the regular pentagon are based on construction of a Golden Section.

The figure at the top of page 1 shows a regular pentagon inscribed in a circle. Two diagonals, $\overline{AC}$ and $\overline{CE}$, are drawn to form a triangle, $\triangle ACE$. The triangle is isosceles because the diagonals of a regular pentagon are congruent. What is the measure of the angle formed at C? (It's a 36° angle.) Can you justify the answer? (The pentagon separates a circle into five congruent arcs, each 72°. Because the measure of an angle inscribed in a circle equals one-half the measure of its intercepted arc, $\angle C$ measures (72/2)° or 36°.) Do you see that $\triangle ACE$ is a Golden Triangle?

When you draw the other diagonals of the pentagon on page 1, a star called a *pentagram* is formed. Also formed are four more Golden Triangles, $\triangle BDA$, $\triangle CEB$, $\triangle DAC$, and $\triangle EBD$. How many different occurrences of the Golden Ratio can you find in these triangles?

First, look at $\triangle ACE$ again. Suppose $CE = 1$. Then, because $\triangle ACE$ is Golden, CE/EA forms the Golden Ratio. ($CE/EA = 1/(1/\phi) = \phi$ where ϕ stands for the Golden Ratio.) Similarly, $CA/EA = \phi$. That's two occurrences of the Golden Ratio in $\triangle ACE$. Because all five of the Golden Triangles are congruent, there are two similar occurrences of ϕ for each one—a total of ten occurrences of the Golden Ratio. But, if $CE/EA = \phi$, so also does CE/AB, CE/BC, CE/CD, and CE/DE. This result gives four more Golden Ratios for each diagonal. How many occurrences in all? Be careful not to count any twice. (There are 25 relationships suggested.)

There are two other sets of Golden Triangles in the pentagon star. Triangle AEI is a Golden Triangle. Can you name all the triangles congruent to $\triangle AEI$? ($\triangle EFA$, $\triangle BAJ$, $\triangle AGB$, $\triangle CBF$, $\triangle BHC$, $\triangle DCG$, $\triangle CID$, $\triangle EDH$, $\triangle DJE$.)

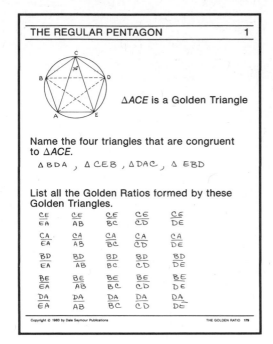

THE REGULAR PENTAGON   1

$\triangle ACE$ is a Golden Triangle

Name the four triangles that are congruent to $\triangle ACE$.

$\triangle BDA$, $\triangle CEB$, $\triangle DAC$, $\triangle EBD$

List all the Golden Ratios formed by these Golden Triangles.

$\frac{CE}{EA}$	$\frac{CE}{AB}$	$\frac{CE}{BC}$	$\frac{CE}{CD}$	$\frac{CE}{DE}$
$\frac{CA}{EA}$	$\frac{CA}{AB}$	$\frac{CA}{BC}$	$\frac{CA}{CD}$	$\frac{CA}{DE}$
$\frac{BD}{EA}$	$\frac{BD}{AB}$	$\frac{BD}{BC}$	$\frac{BD}{CD}$	$\frac{BD}{DE}$
$\frac{BE}{EA}$	$\frac{BE}{AB}$	$\frac{BE}{BC}$	$\frac{BE}{CD}$	$\frac{BE}{DE}$
$\frac{DA}{EA}$	$\frac{DA}{AB}$	$\frac{DA}{BC}$	$\frac{DA}{CD}$	$\frac{DA}{DE}$

THE GOLDEN RATIO 179

How many triangles? (10.) Can you show that $AE/EI = \phi$ and $AI/IE = \phi$? In total, there are 20 such occurrences of the Golden Ratio in this set of Golden Triangles. Can you list them all? (AE/EI, AI/IE, EA/AF, EF/FA, BA/AJ, BJ/JA, AG/BG, AB/BG, CB/BF, CF/FB, BC/CH, BH/HC, DC/CG, DG/GC, CD/DI, CI/ID, ED/DH, EH/HD, DE/EJ, DJ/JE.) In all, the ten triangles congruent to $\triangle AEI$ form 150 different Golden Ratios, 10 ratios for each "isosceles" leg of the triangles.

Each "point" of the pentagram is a Golden Triangle. For example, $\triangle CHG$ is Golden. These five "point" triangles give 50 more occurrences of ϕ. (CH/GH, CH/GF, ... give five. CG/GH, CG/GF, ... give five more, and so on. The total is 10×5 or 50.)

Lest you think the list of Golden Triangles and Golden Ratios is exhausted, consider the segments of each diagonal. The five diagonals intersect each other in five different points—F, G, H, I, and J. For example, diagonal $\overline{CE}$ is intersected at points H and I. These two intersections generate six occurrences of the Golden Ratio. They are CH/HI, CI/IE, EC/CI, EI/IH, IC/CH, and HE/EI. Which of these occurrences have already been counted? (CH/HI and EI/IH were counted when looking at points of the pentagram. CI/IE, IC/CH, and HE/EI were counted when looking at the triangles congruent to $\triangle AEI$.) How many new Golden Ratios can be counted by considering each diagonal? (Ten per diagonal for 50 new ratios. For example, EC/CI, EC/CF, EC/DJ, CE/DG, CE/EF, CE/EH, CE/AI, CE/AG, CE/BJ, and CE/BH.)

In all, you should have counted 20 distinct Golden Triangles and 275 Golden Ratios ($25 + 150 + 50 + 50$). Do you think you've counted them all? You might say that the pentagon star is a gold mine! Just imagine how many more you could find if you drew another pentagram inside the small pentagon.

The following chain of reasoning provides sufficient information for you to easily verify all of the statements found in the discussion of the regular pentagon. Fill in the blanks with the correct information.

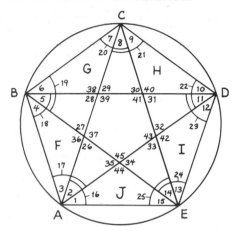

1. Because congruent chords of a circle cut off congruent arcs, each arc formed by the regular pentagon has a measure of 72°. Because inscribed angles of a circle are measured by one-half of their intercepted arcs, Angles 1–15 each have a measure of ____ and Angles 16–25 each have a measure of ____.

1. 36, 72

2. Also, an angle formed by two intersecting chords is measured by one-half the sum of its intercepted arc and the intercepted arc of its vertical angle. Therefore, Angles 26–35 each have a measure of $(1/2)$(____ + ____) or ____. Angles 36–45 each have a measure of $(1/2)$(____ + ____) or ____.

2. 72, 72, 72, 72, 144, 108

3. By the definition of regular pentagon, segments $\overline{AB}$, $\overline{BC}$, ____, ____, and ____ are congruent.

3. $\overline{CD}, \overline{DE}, \overline{EA}$

4. It follows that by the angle-side-angle postulate for congruent triangles, triangles *AFB*, ____, ____, ____, and ____ are congruent.

4. *BGC, CHD, DIE, EJA*

5. Because corresponding sides of congruent triangles are congruent and two sides opposite two congruent angles in a triangle are congruent, $\overline{AF} = \overline{FB} = \overline{BG} = \overline{GC} =$ ____ = ____ = ____ = ____ = ____ = ____.

5. $\overline{CH}, \overline{HD}, \overline{DI}, \overline{IE}, \overline{EJ}, \overline{JA}$

6. As a result, by the side-angle-side postulate for congruent triangles, triangles JAF, FBG, ___, ___, and ___ are congruent.

6. GCH, HDI, IEJ

7. Corresponding parts of congruent triangles are congruent implies that segments $\overline{JF}$, $\overline{FG}$, ___, ___, and ___ are congruent.

7. $\overline{GH}$, $\overline{HI}$, $\overline{IJ}$

8. Therefore, pentagon $FGHIJ$ is equilateral and, because all of its angles have measures of ___, it is equiangular. Consequently, it is a regular pentagon.

8. $108°$

9. If you draw in diagonals $\overline{GJ}$, ___, ___, ___, and ___, this new regular pentagon and pentagram will contain ___ Golden Triangles and ___ Golden Ratios.

9. $\overline{GI}$, $\overline{HF}$, $\overline{HJ}$, $\overline{IF}$, 20, 275

A SPECIAL TRIANGLE

There are many occasions in mathematics when it is helpful to find a *range* of values that an unknown can take. For example, think about a triangle whose sides measure 1, *r*, and r^2. What values can *r* take? Is any real number possible? Or are there limits to the possible values of *r*?

To find the range of values for *r*, consider two different cases.

CASE 1: $r \geq 1$

By the Triangle Inequality, $1 + r > r^2$. In other words, $r^2 - r - 1 < 0$. The solutions to $r^2 - r - 1 = 0$ are $(1 + \sqrt{5})/2$ and $(1 - \sqrt{5})/2$, so you can write the following inequality.

$$\left(r - \frac{1 + \sqrt{5}}{2}\right)\left(r - \frac{1 - \sqrt{5}}{2}\right) < 0$$

If the product of two real numbers is negative, then one of the numbers is negative and the other number is positive. Using this simple property of numbers and remembering that $r \geq 1$, you can say that *r* is limited in the following way.

$$1 \leq r < \frac{1 + \sqrt{5}}{2}$$

CASE 2: $0 < r < 1$

The Triangle Inequality gives you $r^2 + r > 1$. (Note that this triangle *differs* from the triangle in Case 1.) Using the Triangle Inequality and an argument similar to the one used in Case 2, you can find the following inequality.

$$\left(r - \frac{-1 + \sqrt{5}}{2}\right)\left(r - \frac{-1 - \sqrt{5}}{2}\right) > 0$$

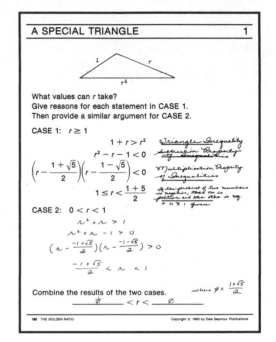

A SPECIAL TRIANGLE   1

What values can *r* take?
Give reasons for each statement in CASE 1.
Then provide a similar argument for CASE 2.

CASE 1: $r \geq 1$

$1 + r > r^2$ *Triangle Inequality*
$r^2 - r - 1 < 0$ *Subtraction Property of Inequalities*
$\left(r - \frac{1 + \sqrt{5}}{2}\right)\left(r - \frac{1 - \sqrt{5}}{2}\right) < 0$ *Multiplication Property of Inequalities*
$1 \leq r < \frac{1 + 5}{2}$ *If the product of two numbers is negative, then one is positive and the other is neg. $+ r \geq 1$ given*

CASE 2: $0 < r < 1$

$r^2 + r > 1$
$r^2 + r - 1 > 0$
$\left(r - \frac{-1 + \sqrt{5}}{2}\right)\left(r - \frac{-1 - \sqrt{5}}{2}\right) > 0$
$\frac{-1 + \sqrt{5}}{2} < r < 1$

Combine the results of the two cases.   *where $\phi = \frac{1 + \sqrt{5}}{2}$*
$\frac{1}{\phi} < r < \phi$

182 THE GOLDEN RATIO   Copyright © 1983 by Dale Seymour Publications

For the product of two real numbers to be positive, they must both be positive or both negative. This fact, along with the condition $0 < r < 1$, results in the following inequality.

$$\frac{-1 + \sqrt{5}}{2} < r < 1$$

COMBINED RESULTS:

Combining the results of the two cases gives a very interesting inequality.

$$\frac{-1 + \sqrt{5}}{2} < r < \frac{1 + \sqrt{5}}{2} \qquad \text{or} \qquad \frac{1}{\phi} < r < \phi$$

where ϕ stands for the Golden Ratio

That is, if a triangle has sides whose measures are in geometric progression—1, r, r^2—then the value of r is between the value of the Golden Ratio and its reciprocal.

EXERCISES

1. How does the value of r for a triangle whose sides measure a, ar, and ar^2 where a is positive (any triangle whose sides have measures that form a geometric progression) compare to the value of r when $a = 1$?

ANSWERS

1. The factor of a will either increase or decrease the measurement of the sides, but in either case it will produce a triangle with sides proportional to those of a triangle with $a = 1$. The value of a does *not* affect the range of r because it divides out in each case.

   1. $r \geq 1$; $a + ar > ar^2$
   2. $0 < r < 1$; $ar^2 + ar > a$

2. Each threesome of numbers corresponds to the measures of the sides of a triangle. Tell whether or not the measures are in geometric progression, and if they are, give the ratio r.

sides of triangle	geometric?	r
1, 0.8, 0.64		
3, 2.4, 1.92		
4, 14, 16		
1/3, 1/4, 3/16		
1, 1, 1		
2, 1.4, 0.98		
7, 7, 10		
5, 5 $\sqrt{2}$, 10		

3. Find the solution set for $x^2 + 2x - 15 > 0$.

4. Find the solution set for $x^2 + 9x + 18 < 0$.

2. yes, 0.8; yes, 0.8; no; yes, 3/4; yes, 1; yes, 0.7; no; yes, $\sqrt{2}$

3. The set of all real numbers x such that $x < -5$ or $x > 3$

4. The set of all real numbers x such that $-6 < x < -3$

THE CENTRAL ANGLE OF A PENTAGON

MATHEMATICS CONTENT:
Ratio,
Radical,
Cosine,
Golden Section

The Golden Ratio often appears when you least expect it. Consider, for example, the central angle of a regular pentagon (formed by connecting the center to two adjacent vertices). There are five central angles in a pentagon, all congruent to each other. If the pentagon is inscribed in a circle, each central angle will intercept an arc measuring $(360/5)°$ or $72°$. Thus, each central angle measures $72°$ (because a central angle is measured by its intercepted arc).

Now use a calculator or a table of cosines to find a value for cos 72°. (Your value should be about 0.309017.) Next, find a decimal approximation for $1/2\phi$ where ϕ stands for the Golden Ratio. Then compare your two answers.

Can you prove that, in fact, $\cos 72° = 1/(2\phi)$?

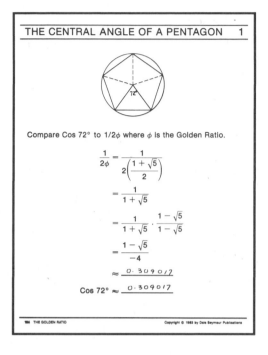

THE CENTRAL ANGLE OF A PENTAGON   1

Compare Cos 72° to 1/2φ where φ Is the Golden Ratio.

$$\frac{1}{2\phi} = \frac{1}{2\left(\dfrac{1 + \sqrt{5}}{2}\right)}$$

$$= \frac{1}{1 + \sqrt{5}}$$

$$= \frac{1}{1 + \sqrt{5}} \cdot \frac{1 - \sqrt{5}}{1 - \sqrt{5}}$$

$$= \frac{1 - \sqrt{5}}{-4}$$

$$\approx 0.309017$$

$$\text{Cos } 72° \approx 0.309017$$

THE GOLDEN RATIO   Copyright © 1983 by Dale Seymour Publications

EXERCISES

1.–4. The discussion of the pentagon star inscribed in a circle gave many examples of ratios of line segments that are Golden Ratios. What about the angles of the figure in this lesson? (Connect all vertices to the center.) Are they in some way related to the Golden Ratio? Answer this question for yourself by completing the following table. (Note that the angle measuring 144° is supplementary to an angle measuring 36°, so it is included in this table for your interest.)

degree measure of angle	Cosine decimal form	Cosine radical form	Cosine φ form
72			
36			
108			
144			

ANSWERS

1. 72° angle: 0.309017,
   $1/[2(1 + \sqrt{5})/(2)]$, $1/2\phi$

2. 36° angle: 0.809017,
   $[(1 + \sqrt{5})/2]/2$, $\phi/2$

3. 108° angle: -0.309017,
   $-(1/[2(1 + \sqrt{5})/2])$, $-(1/2\phi)$

4. 144° angle: -0.809017,
   $-([(1 + \sqrt{5})/2]/2)$, $-(\phi/2)$

NATURE'S PENTAGRAMS

What do flower blossoms have to do with the Golden Ratio? Have you ever noticed that five-petaled flower blossoms have the shape of a pentagram? If you measure the distance from the tip of one petal to the tip of a *nonadjacent* petal and then divide that distance by the distance between two *adjacent* petal tips, you will get an approximation of the Golden Ratio.

There are hundreds of flower species with pentagram blossoms. In fact, more flower blossoms have five petals than any other number of petals. The list on page 1 names only a few of the many kinds of pentagram blossoms.

Some flower blossoms look like a star on a star. A blossom will have two pentagrams of petals, one on top of the other, or a blossom will have five petals atop five sepals. Two very striking examples are the hoya plant blossom and the passion plant blossom. The hoya plant, found mainly in the region from Australia and the New Hebrides to Western Polynesia, is grown as a house plant in the continental United States. Students in my classes admire the beautiful geometric design of the star on a star in hoya blossoms from my own plant. Passion flowers grow in Hawaii and on the coast of California. I once saw a purple passion flower in a restaurant garden on the Berkeley campus of the University of California.

Once you become aware of nature's pentagrams, you see them everywhere—from under the sea to the inside of your own refrigerator. Besides the hundreds of species of five-petaled flowers, nature displays pentagram designs in many parts of the plant and animal kingdoms. The starfish, the cross section of an apple seedbed, and the sand dollar are three such examples. Can you name more?

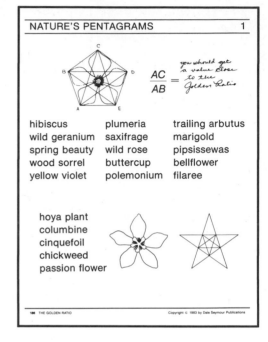

NATURE'S PENTAGRAMS 1

$$\frac{AC}{AB} =$$ *you should get a value close to the Golden Ratio*

hibiscus   plumeria   trailing arbutus
wild geranium saxifrage   marigold
spring beauty wild rose   pipsissewas
wood sorrel   buttercup   bellflower
yellow violet   polemonium filaree

hoya plant
columbine
cinquefoil
chickweed
passion flower

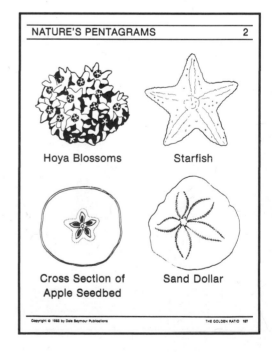

NATURE'S PENTAGRAMS 2

Hoya Blossoms   Starfish

Cross Section of
Apple Seedbed   Sand Dollar

EXERCISES

1. Find the names of ten flowers whose blossoms contain five petals and are *not* on the list in this lesson.

2. Name a plant that has either five petals on five additional petals or five petals on five sepals and does *not* appear on the list in this lesson.

3. Name two examples of pentagrams in nature that are *not* mentioned in this lesson.

ANSWERS

1. Answers will vary.

2. Answers will vary.

3. Answers will vary.

SIX

THE 13TH CENTURY STILL LIVES

There is good evidence to support the belief that Egyptians and Greeks long ago knew about a sequence of numbers called Fibonacci numbers, but it wasn't until the 13th century that interest in the numbers flowered. Early in that century Leonardo da Pisa, also known as Leonardo Fibonacci, published *Liber Abaci* (The Book of the Abacus) in which he posed a problem about rabbits.

> A pair of rabbits one month old is too young to produce more rabbits. Suppose that in their second month they produce a new pair. If each new pair of rabbits does the same, and if none of the rabbits die, how many pairs of rabbits will there be at the beginning of each month?

The solution to Fibonacci's rabbit problem gives rise to the sequence of numbers still named after him—the Fibonacci sequence.

The occurrences of Fibonacci numbers within the biological and physical worlds are so widespread that there is an entire organization in the United States devoted solely to their study—The Fibonacci Association. Since 1963, the association has published a quarterly journal, *The Fibonacci Quarterly,* that is filled with information about the famous numbers. The lessons in this book can only give a brief introduction to these fascinating numbers—numbers that are some of the most common numbers of life itself.

The information in the lessons that follow are based on ideas from:

Mathematics by David Bergamini and the editors of Time-Life Books, pages 92 and 93.

"Fibonacci Sequences" by Brother Alfred Brousseau in *Topics for Mathematics Clubs,* pages 5–8.

Fibonacci and Lucas Numbers by Verner E. Hoggatt, Jr., pages 26–29, 48–50, and 79–82.

The Divine Proportion: A Study in Mathematical Beauty by H. E. Huntley, pages 49–50, 131–133, 145, 148, and 161–165.

The Golden Section and Related Curiosa by Garth E. Runion, pages 71–73 and 91–96.

The third edition of *Using Advanced Algebra* by K. J. Travers, L. C. Dalton, and V. F. Brunner, pages 425 and 429.

FIBONACCI, RABBITS, AND FLOWERS

If you solve Fibonacci's rabbit problem, you will find that there is 1 pair of rabbits at the beginning of the first month, 1 rabbit pair the second month, 2 pairs the third month, 3 pairs the fourth month, 5 pairs the next, and so on, to obtain the following sequence.

$$1, 1, 2, 3, 5, 8, 13, 21, 34, \ldots$$

What are the next few terms? Do you see the pattern? After the first two months, you add the terms from the previous two months to find the next term.

$$1 + 1 = 2, \quad 1 + 2 = 3, \quad 2 + 3 = 5, \quad 3 + 5 = 8, \ldots$$

It is this simple property that defines the Fibonacci sequence.

Many flower species are partial to Fibonacci numbers. Their petals commonly occur only in Fibonacci number configurations. Enchanter's Nightshade flowers, for example, have *two* petals. Trilium and lilies have *three* petals. The most prevalent Fibonacci number among flower petals is *five*. There are, however, flowers that exhibit even greater numbers of petals. Page 1 lists some flowers and their Fibonacci petal numbers. How many petals do field daisies usually have? What about Michaelmas daisies?

Sunflowers are very special flowers when it comes to the Fibonacci sequence. The seeds in their flower heads spiral in two different directions. Look at the picture on page 2. Count the spirals in each direction. (There are 34 clockwise and 21 counterclockwise.) The number of spirals is a Fibonacci number, usually 34 spirals one way and 55 spirals the other. The table on page 2 shows some commonly occurring pairs.

Like sunflower seeds, pineapple scales and pine cone scales spiral in two different directions and the numbers of spirals are Fibonacci numbers. Pineapples usually have 8

FIBONACCI, RABBITS, AND FLOWERS 1

1, 1, 2, 3, 5, 8, 13, 21, 34, _55_ , _89_ , . . .

Number of Petals	Name of Flower	Number of Petals	Name of Flower
2	Enchanter's Nightshade	21	Chicory Aster Helenium
3	Trilium Lily Iris	34	Plantain Ox-eye Daisy Pyrethrum
5	Wild Geranium Spring Beauty Yellow Violet	_55_	Field Daisy Helenium Michaelmas Daisy
8	Lesser Celandine Sticktight Delphinium	_89_	Michaelmas Daisy
13	Corn Marigold Mayweed Ragwort		

FIBONACCI, RABBITS, AND FLOWERS 2

SUNFLOWER SPIRALS

one way	8	21	34	55	89
other way	13	34	55	89	144

PINEAPPLE SPIRALS	PINE CONE SPIRALS
5 and 8, 8 and 13	5 and 8

34 and 21

How many spirals?

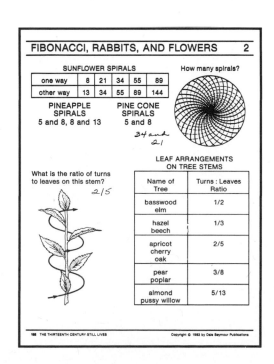

What is the ratio of turns to leaves on this stem? 2/5

LEAF ARRANGEMENTS ON TREE STEMS

Name of Tree	Turns : Leaves Ratio
basswood elm	1/2
hazel beech	1/3
apricot cherry oak	2/5
pear poplar	3/8
almond pussy willow	5/13

spirals in one direction and 13 in the other. For pine cones, there are usually 5 spirals winding one way and 8 spirals winding the other.

In many trees, the leaves spiral around the stems. The number of turns required to find a leaf in a position *directly above* another leaf is a Fibonacci number. In addition, the number of leaves within those turns is a Fibonacci number. For example, in a pear tree there are usually 3 turns around a stem before a leaf is directly above another leaf and there are normally 8 leaves within those 3 turns (only count one of the two leaves directly above and below one another). My own students have counted leaves and turns from my back-yard pear tree, and each time they have found this 3 turns/8 leaves configuration. Can you find the turns/leaves ratio in the picture? (There are 2 turns/5 leaves.)

EXERCISES

1. In Fibonacci's rabbit problem, how many pairs of rabbits will there be at the beginning of the seventh month? How many pairs will there be at the beginning of the twelfth month?

2. In Fibonacci's rabbit problem, how many pairs of adult rabbits (at least one month old) will there be at the beginning of the seventh month? How many baby rabbits (less than one month old) will there be?

3. Solve Exercise 2 for the beginning of the twelfth month.

4. Fibonacci numbers have some remarkable properties. Find the missing numbers in this sequence of sums. Describe the pattern.

$$1^2 = 1 \times 1$$
$$1^2 + 1^2 = 1 \times 2$$
$$1^2 + 1^2 + 2^2 = 2 \times 3$$
$$1^2 + 1^2 + 2^2 + 3^2 = 3 \times \underline{}$$
$$1^2 + 1^2 + 2^2 + 3^2 + 5^2 = \underline{} \times \underline{}$$
$$1^2 + 1^2 + 2^2 + 3^2 + 5^2 + 8^2 = \underline{} \times \underline{}$$
$$1^2 + 1^2 + 2^2 + 3^2 + 5^2 + 8^2 + 13^2 = \underline{} \times \underline{}$$

ANSWERS

1. 21, 233

2. 13, 8

3. 144, 89

4. 3×5, 5×8, 8×13, 13×21; Each sum of consecutive Fibonacci squares equals a product of two consecutive Fibonacci numbers. The two factors are the last Fibonacci number square in the sum and its successor. The sequences of factors create two Fibonacci sequence "columns" except for a missing 1 in one of the columns.

5. Find examples of at least 5 flowers named in this lesson. Find pictures from a book about flowers. Then try to find these flowers in gardens, along roadsides, or in parks. Count the petals. Are your results Fibonacci numbers?

5. Answers will vary.

6. Count the spirals, clockwise and counterclockwise, in a real pineapple. Are there 8 spirals one way and 13 the other?

6. Answers will vary.

7. Verify at least one *turns-to-leaves* ratio from the stem of a real tree that is listed in this lesson.

7. Answers will vary.

FIBONACCI NUMBERS
AND THE GOLDEN RATIO

MATHEMATICS CONTENT:
Fibonacci Sequence,
Golden Ratio,
Ratio, Limit of Sequence,
Geometric Sequence,
Additive Property

If you divide each term of the Fibonacci sequence by its preceding term and look at the sequence of numbers you obtain, you will make an interesting discovery. Page 1 gives the values for the first few ratios. Use a calculator to help you determine the ratios for values of *n* up to 15. What do you see? (The sequences of numbers get closer and closer together. The numbers appear to approach a certain value.) Now find a decimal approximation for $(1 + \sqrt{5})/2$, the Golden Ratio. (Correct to six decimal places, the approximation is 1.618034.) Do you see how close the terms of your sequence get to the Golden Ratio? In fact, if you continue the dividing process, you can get as close as you wish to any known approximation of $(1 + \sqrt{5})/2$.

Mathematicians use the following expression to describe the situation.

$$\lim_{n \to \infty} \frac{F_{n+1}}{F_n} = \frac{1 + \sqrt{5}}{2}$$

In this expression, F_n and F_{n+1} stand for terms of the Fibonacci sequence, *n* and *n* + 1 representing consecutive positions of the terms in the sequence.

As you might suspect, the Golden Ratio and the Fibonacci sequence are related in several other ways. A very special geometric sequence is formed when you take powers of the Golden Ratio, ϕ.

$$\phi^1, \phi^2, \phi^3, \phi^4, \phi^5, \ldots$$

Decimal approximations for the first few terms of the sequence are 1.618034, 2.618034, 4.236068, 6.8541020, and 11.090170. A closer inspection of the terms tells you that the sequence of powers is equivalent to the following sequence.

$$1\phi + 0, 1\phi + 1, 2\phi + 1, 3\phi + 2, 5\phi + 3, \ldots$$

Can you name the next three terms of this sequence? Find decimal approximations for your terms and the terms ϕ^6, ϕ^7, and ϕ^8 to check your answers.

**FIBONACCI NUMBERS &
THE GOLDEN RATIO** 1

n	$\dfrac{F_{n+1}}{F_n}$	Decimal Approximation
1	1/1	1.00000000
2	2/1	2.00000000
3	3/2	1.50000000
4	5/3	1.66666667
5	8/5	1.60000000
6	13/8	1.62500000
7	21/13	1.61538462
8	34/21	1.61904762
9	55/34	1.61764706
10	89/55	1.61818182
11	144/89	1.61797753
12	233/144	1.61805556
13	377/233	1.61802575
14	610/377	1.61803714
15	987/610	1.61803279

$$\lim_{n \to \infty} \frac{F_{n+1}}{F_n} = \frac{1 + \sqrt{5}}{2}$$
$$\approx 1.61803399$$

**FIBONACCI NUMBERS &
THE GOLDEN RATIO** 2

$\phi^1 \approx 1.6180340 \approx 1\phi + 0$

$\phi^2 \approx 2.6180340 \approx 1\phi + 1$

$\phi^3 \approx 4.2360680 \approx 2\phi + 1$

$\phi^4 \approx 6.854120 \approx 3\phi + 2$

$\phi^5 \approx 11.090170 \approx 5\phi + 3$

$\phi^6 \approx 17.944272 \approx 8\phi + 5$

$\phi^7 \approx 29.034442 \approx 13\phi + 8$

$\phi^8 \approx 46.928714 \approx 21\phi + 13$

The numbers in the new form of the sequence should remind you of the Fibonacci sequence. You should also see that this sequence has the "additive" property of the Fibonacci sequence; that is, if you add any two consecutive terms, you will obtain the very next term of the sequence. For example, $(3\phi + 2) + (5\phi + 3) = (8\phi + 5)$. So this special sequence is both geometric and additive!

EXERCISES

1. Continue to find ratios F_{n+1}/F_n until you obtain an approximation for ϕ that is correct to six decimal places (1.618034). What is the least value of n that gives an approximation of ϕ correct to six decimal places?

2. In a geometric sequence, each term after the first term can be determined by multiplying the previous term by a constant. Normally, a geometric sequence is represented by $a_1, a_2, a_3, \ldots$ and r stands for the constant. In general, the terms of a geometric sequence can be described by the formula $a_n = a_1 r^{n-1}$ where a_n represents the nth term. What are the values of a_1 and r in the sequence $\phi^1, \phi^2, \phi^3, \phi^4, \ldots$?

3. Give an expression for the general term of the sequence $\phi^1, \phi^2, \phi^3, \phi^4, \ldots$ as it would be described using the formula in Exercise 2. Then use the formula to generate the first three terms of the sequence.

4. Use $(1 + \sqrt{5})/2$ for ϕ and the values of Fibonacci numbers F_1 and F_2 to show that $\phi^2 = (F_2)\phi + F_1$.

5. Use $(1 + \sqrt{5})/2$ for ϕ and the values of Fibonacci numbers F_2 and F_3 to show that $\phi^3 = (F_3)\phi + F_2$.

ANSWERS

1. $F_{16}/F_{15} = 987/610 \approx$ 1.6180328; $F_{17}/F_{16} =$ 1597/987 $\approx$ 1.6180344; the least value of n is 16.

2. $a_1 = \phi; r = \phi$

3. $a_n = \phi(\phi)^{n-1}$;
$a_1 = \phi(\phi)^{1-1} = \phi^1$;
$a_2 = \phi(\phi)^{2-1} = \phi^2$;
$a_3 = \phi(\phi)^{3-1} = \phi^3$

4. $\phi^2 = [(1 + \sqrt{5})/2]^2$
$= (6 + 2\sqrt{5})/4 = (3 + \sqrt{5})/2$
$= [(1 + \sqrt{5})/2 + (2/2)]$
$= [(1 + \sqrt{5})/2 + (1)]$
$= (1)\phi + 1 = F_2(\phi) + F_1$

5. $\phi^3 = [(1 + \sqrt{5})/2]^3$
$= [(1 + \sqrt{5})/2]^2[(1 + \sqrt{5})/2]$
$= [(3 + \sqrt{5})/2] \times$
$\qquad\qquad [(1 + \sqrt{5})/2]$
$= (8 + 4\sqrt{5})/4 = 2 + \sqrt{5}$
$= 2[(1 + \sqrt{5})/2] + 1$
$= F_3(\phi) + F_2$

BINET'S FORMULA

Having an expression for the general term (the nth term) of a sequence often makes the job of finding a specific term of the sequence much easier—especially if you're interested in the value of the 100th term or the 1000th term. There is an expression for the general term of the Fibonacci sequence known as Binet's Formula. Would it surprise you to know that the Golden Ratio is part of this general term? Both solutions to the Golden Quadratic Equation ($x^2 - x - 1 = 0$) are part of Binet's Formula. If F_n represents the value of the general term (the nth term) of the Fibonacci sequence, then Binet's Formula is as follows.

$$F_n = \frac{1}{\sqrt{5}}\left(\frac{1 + \sqrt{5}}{2}\right)^n - \frac{1}{\sqrt{5}}\left(\frac{1 - \sqrt{5}}{2}\right)^n$$

Binet's Formula gives you $(1/\sqrt{5})[(1 + \sqrt{5})/2]^{10} - (1/\sqrt{5})[(1 - \sqrt{5})/2]^{10}$ for the tenth term of the Fibonacci sequence. With the help of a scientific calculator, you can quickly obtain a simplified value. The exact value is 55. Some scientific calculators will give you this value. Some will be off by a very small amount.

Without a calculator, your work is more tedious. Try, for example, to find the value of F_2 from Binet's Formula without using a calculator. For powers greater than 2, you would probably need to use both logarithms and the binomial theorem to help.

Calculators do have their limitations. The values you obtain on your calculator using Binet's Formula are only approximate. Some approximations are so close that it appears you have an exact answer. (For example, a TI-35 calculator will give an apparently exact value for F_{10}.) However, as the values of n increase, the approximations you get are less and less accurate. F_{100} has so many digits that it won't even fit on a calculator display. The exact value of F_{100} is 354,224,848,-179,261,915,075. Two different calculators gave approximations of 3.5422×10^{20} and 3.5422484×10^{20}. What approximation will you get for F_{100}?

BINET'S FORMULA   1

Binet's Formula:

$$F_n = \frac{1}{\sqrt{5}}\left(\frac{1 + \sqrt{5}}{2}\right)^n - \frac{1}{\sqrt{5}}\left(\frac{1 - \sqrt{5}}{2}\right)^n$$

Use a calculator and Binet's formula to find F_{10}.

Answers will vary.

$F_{10} = 55$

Without the aid of a calculator, use Binet's formula to find F_2.

$$F_2 = \frac{1}{\sqrt{5}}\left(\frac{1 + \sqrt{5}}{2}\right)^2 - \frac{1}{\sqrt{5}}\left(\frac{1 - \sqrt{5}}{2}\right)^2$$

$$= \frac{1}{\sqrt{5}}\left(\frac{6 + 2\sqrt{5}}{4}\right) - \frac{1}{\sqrt{5}}\left(\frac{6 - 2\sqrt{5}}{4}\right)$$

$$= \frac{1}{\sqrt{5}}\left(\frac{6 + 2\sqrt{5} - 6 + 2\sqrt{5}}{4}\right)$$

$$= \frac{\sqrt{5}}{\sqrt{5}}$$

$$= 1$$

EXERCISES

1. Without the aid of a calculator, use Binet's Formula to find F_3.

2. Without the aid of a calculator, use Binet's Formula to find F_4.

3. Use a calculator and Binet's Formula to find F_{11}.

4. Use a calculator and Binet's Formula to find F_{20}.

ANSWERS

1. See student work.

2. See student work.

3. Answers may vary. $F_{11} = 89$.

4. Answers may vary. $F_{20} = 6765$.

THE CHINESE TRIANGLE

Are you familiar with a numerical pattern called Pascal's Triangle? Page 1 shows one version of this famous pattern. Would you guess that the Fibonacci sequence is related to Pascal's Triangle? Add the numbers along each of the diagonals drawn. What do you get? (The numbers form the Fibonacci sequence.)

Today this array of numbers on page 1 is called Pascal's Triangle because Pascal, in the 1600's, developed many of the number relationships in the triangle and applied them to many different situations. Pascal was not the first to work with the triangle, though. Chinese mathematicians knew about and wrote about the triangle in the thirteenth century. In fact, their writings imply that the triangle had already been known for a long time.

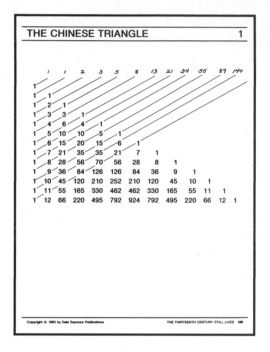

EXERCISES

1. Expand each of the following binomials and write the resulting polynomials in the form of the Chinese triangle. (Use 1's where you see no numerical coefficients.)

   a. $(a + b)^0$
   b. $(a + b)^1$
   c. $(a + b)^2$
   d. $(a + b)^3$
   e. $(a + b)^4$
   f. $(a + b)^5$

ANSWERS

1a. 1
  b. $1a + 1b$
  c. $1a^2 + 2ab + 1b^2$
  d. $1a^3 + 3a^2b + 3ab^2 + 1b^3$
  e. $1a^4 + 4a^3b^1 +$
           $6a^2b^2 + 4a^1b^3 + 1b^4$
  f. $1a^5 + 5a^4b^1 + 10a^3b^2 +$
           $10a^2b^3 + 5a^1b^4 + 1b^5$

2. After you have found all the expansions in Exercise 1, draw diagonals through them as was done with the Chinese triangle in this lesson. Describe the sums of the binomial coefficients obtained by adding along your diagonals.

2. The sums of the binomial coefficients along the diagonals drawn generate the Fibonacci sequence.

3. Find the sum of the numbers in each row of the Chinese triangle for the first ten rows. Describe the sequence of sums.

3. The row sums can be written as consecutive positive integer powers of 2.

SEVEN

MATHEMATICS AND MUSIC*

Two apparently distant and different topics are sometimes closely related. Such is the case with mathematics and music. In searching for the number essence of all things, the Pythagoreans discovered several simple numerical relationships for musical sounds. During the eighteenth century, mathematicians went beyond the Pythagoreans' understanding of music, completing a quite thorough mathematical description and analysis of musical sound. A knowledge of the mathematics of music is fundamental for designing telephones, radios, record players, and televisions.

One of the interesting relationships between mathematics and music shows up in the forms of musical instruments. The spacing of frets on a guitar, the curve of a grand piano, and the size of organ pipes all mirror the mathematics of sound. The lessons that follow show how "sound" mathematics is connected to the shape of pianos, guitars, and organs. Then the lessons take the mathematics further, showing how it relates not only to music but a variety of natural forms. The final lesson, *Coxeter's Golden Sequence,* brings together in one elegant example musical math, the exponential function, the golden ratio, and the logarithmic spiral.

*The ideas that follow are based on information from the following references.

"Logs, Pianos and Spirals" in *The Language of Mathematics* by Frank Land, pages 128–132 and 139–142.

Mathematics: A Human Endeavor by Harold R. Jacobs, pp. 176, 271–273, 284–285, and 290–291.

The Divine Proportion: A Study in Mathematical Beauty by H. E. Huntley, pages 100–102.

The *Teacher's Guide for Math in Nature Posters* by Alan Hoffer, page 10.

Patterns in Nature by Stevens, pages 88–89.

Mathematics, Its Magic and Mastery, Second Edition by Aaron Bakst, pages 295–296.

The Golden Section and Related Curiosa by Garth E. Runion, pages 54–55.

"The Diverse Pleasures of Circles that are Tangent to One Another" in the Mathematical Games section of the January 1979 issue of *Scientific American,* pages 24 and 28.

THE GRAND PIANO

Each of the keys of a piano is associated with a string inside the piano. Striking a piano key causes a "hammer" inside the piano to hit the string in the piano. When the hammer hits the string, the string vibrates. The vibrating string creates a musical sound of a certain pitch.

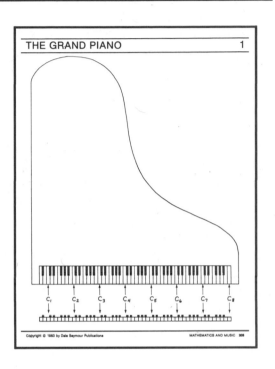

Short strings vibrate very fast, creating high pitches. Long strings vibrate at lesser frequencies, creating lower pitches. A piano is arranged so that, facing the keyboard, the keys give very low pitches at the left and gradually give higher pitches as you move to the right. The string arrangement in a grand piano corresponds to the keyboard arrangement; it has long strings at the left and short strings at the right. Page 1 shows a sketch of a grand piano and its keyboard, viewed from the top. Notice the graceful curve of the piano that accommodates the various lengths of its strings.

If you look carefully at the keyboard of a piano, you will see that the keys are arranged in a pattern. There is a group of seven white keys and five black keys that repeats seven times (with a few keys left over). Each of these groups of keys corresponds to a musical octave.

Notes in musical octaves are related in a very interesting mathematical way. The C's within each octave are marked on the piano of page 1. The *frequencies* of these C-notes double as you go from C to C. (Frequency is a measure of the vibrations of the strings.) The frequency of C_2 is twice the frequency of C_1; the frequency of C_3 is twice the frequency of C_2 and, thus, four times the frequency of C_1; and so on. In other words, the frequencies of these notes are related exponentially and their relationship can be described by the following equation.

$$C_n = C_1 2^{n-1}$$

where $n \in \{1, 2, 3, 4, 5, 6, 7, 8\}$
and C_n represents the frequency
of the nth C on the piano.

Using this equation, you can find the relationship between C_4 and C_1, for example.

$$C_4 = C_1 2^{(4-1)} = C_1 2^3 = 8C_1$$

This means that the frequency of the fourth C on the piano, C_4, is eight times the frequency of the first C on the piano, C_1.

Unless you know the actual frequency of one of the C's on the piano, you cannot find the frequencies of the other C's using the equation. You can, however, find the *relative* frequencies of the various C's. Simply evaluate 2^{n-1} for each value of n. By graphing these relative frequencies for the eight C-notes, you will obtain a visual representation of the relationship among the C's of the piano. Page 2 graphs the first two points. Complete the graph.

The completed graph has eight points. The graph represents an exponential function—an exponential function with a *restricted* domain. What is the domain of your function? (The domain is $\{1, 2, 3, 4, 5, 6, 7, 8\}$.) The function can be described by the equation $f(x) = 2^{x-1}$. If you remove the restriction on the domain and graph $f(x) = 2^{x-1}$ for all real values of x, the result will be an exponential curve. Sketch it in the first quadrant. Each point of your graph lies on this exponential curve. Do you see how the curve is related to the curve of the grand piano?

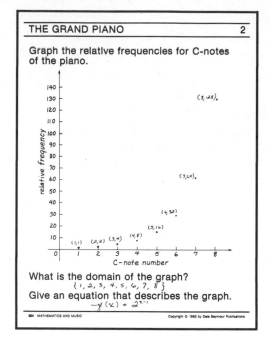

THE GRAND PIANO   2

Graph the relative frequencies for C-notes of the piano.

What is the domain of the graph?
$\{1, 2, 3, 4, 5, 6, 7, 8\}$
Give an equation that describes the graph.
$f(x) = 2^{x-1}$

MATHEMATICS AND MUSIC   Copyright © 1983 by Dale Seymour Publications

EXERCISES

1. Graph $f(x) = 2^x$ where $x \in \{1, 2, 3, 4, 5\}$.

2. Graph $f(x) = 2^x$ where x is a real number.

ANSWERS

1. The points on the graph are $(1, 2)$, $(2, 4)$, $(3, 8)$, $(4, 16)$, and $(5, 32)$.

2. The graph intersects the vertical axis at 1, curves upward to the right and very slowly down to the left; it is asymptotic to the horizontal axis toward the left; it goes through $(-1, 1/2)$ and $(4, 16)$.

3. Graph $f(x) = 3^x$ where x is a real number.

3. The graph has the same basic shape as $f(x) = 2^x$ but rises faster to the right and curves downward to the left more slowly; it intersects the vertical axis at 1; it goes through $(-1, 1/3)$ and $(4, 81)$.

4. Graph $f(x) = 3^{-x}$ where x is a real number.

4. The graph is the reflection of $f(x) = 3^x$ over the vertical axis.

5. Graph $f(x) = 4(2^x)$ where x is a real number.

5. The graph is steeper than the graph of $f(x) = 2^x$; it intersects the vertical axis at 4; it goes through $(1, 8)$ and $(-1, 2)$.

6. Graph $f(x) = 4(2^{-x})$ where x is a real number.

6. The graph is the reflection of the graph in Exercise 5 over the vertical axis.

7. Graph $f(x) = 5(2^x)$ where x is a real number.

7. The graph is steeper than both $f(x) = 2^x$ and $f(x) = 4(2^x)$; it intersects the vertical axis at 5; it goes through $(1, 10)$ and $(-1, 2\text{-}1/2)$.

8. Graph $f(x) = 5(2^{-x})$ where x is a real number.

8. The graph is the reflection of the graph in Exercise 7 over the vertical axis.

MORE PIANO RELATIONSHIPS

Every octave of the piano contains thirteen notes from C to C. There are eight white keys and five black keys—two and three. (Have you seen the numbers 2, 3, 5, 8 and 13 before?) The figure on page 1 shows a close-up of the middle octave, or scale, on a piano keyboard.

Today pianos are usually tuned so that the ratios of frequencies of consecutive notes are constant. For example, the ratio of frequencies from middle C to C# is the same as the ratio of frequencies from C# to D and from D to D# and so on. (The notes in order are C, C#, D, D#, E, F, F#, G, G#, A, A#, B, and C.) Suppose r stands for that ratio and f_n stands for the frequency of the nth note of the scale starting from middle C. Can you write an equation describing how f_n is related to f_1, the frequency of middle C? Recall that the numbers in a geometric sequence are related by a common ratio.

The equation relating the frequencies is like the equation describing the nth term of a geometric sequence. The equation you're looking for is

$$f_n = f_1 r^{n-1}$$

The frequency of middle C is about 261.6 cycles per second. Because the frequencies of C-notes double from octave to octave, the frequency of the C above middle C is about 523.2 cycles per second. Using these two values you can find a value for the common ratio r between notes of the scale. If you don't have a calculator, you can use logarithms to solve the problem. (The value of r is $\sqrt[12]{2}$. You will get log $r = (1/12) \log 2 \approx 0.0251$, so $r \approx 1.0595$. Using a TI-35 calculator, you will find r is approximately 1.0594631.)

Once you've found a value for r, you can find approximate frequencies for all the notes of the middle octave. Then you can graph the frequencies with respect to the numbers you've assigned to the notes. What is the domain of your graph?

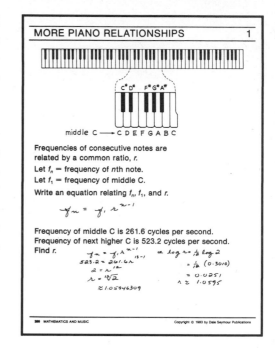

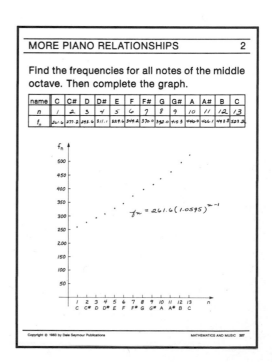

(The domain is 1, 2, 3, 4, 5, 6, 7, 8, 9, 10, 11, 12, 13.) What is the range? (The range is 261.6, 277.2, 293.6, 311.1, 329.6, 349.2, 370.0, 392.0, 415.3, 440.0, 466.1, 493.8, 523.2. These values are approximate and your results may differ by a few tenths.)

EXERCISES

1. Use the equation $f_n = f_1 r^{n-1}$ to verify that f_{13}, the frequency of C above middle C, is 523.2 cycles per second given that f_1 is 261.6 and $r \approx 1.05946$.

2. Describe how to extend the graph of this lesson to show the frequencies of notes in the next octave above the middle octave of a piano.

3. What is the frequency of F in the next octave of a piano?

4. What is the frequency of the upper C in the next octave above the middle octave of a piano? Describe *two* different methods for finding the frequency.

5. How is the frequency of F in the middle octave of a piano related to the frequency of F in the next octave above?

ANSWERS

1. $f_{13} \approx 261.6 \times (1.05946)^{13-1}$
   $\approx 261.6(1.99993) \approx 523.2$.

2. Continue numbering the notes of the next octave consecutively. The first C is the last C of the middle octave and is numbered 13. C# is numbered 14, D is numbered 15, and so on. Then use the equation $f_n = f_1 r^{n-1}$ to find the values of the range. The extended domain is $\{1, 2, \ldots, 25\}$.

3. Approximately 698.4 cycles per second.

4. You can multiply the frequency of middle C by 2^2 or you can find $261.6r^{24}$.

5. The frequencies are related by a factor of r^{12}; that is, by 2.

GUITAR FRETS

Just as the frequencies of notes in a musical scale are related, the wavelengths of those notes are related. In fact, the relationships are very similar. The wavelength of a musical note is exactly twice the wavelength of the note one octave *higher*. (Compare this to the frequency relationship of the two notes.) Also, the wavelengths of consecutive notes within the octave differ by $\sqrt[12]{2}$ or about 1.0595.

The mathematics of sound is reflected in the shape of a grand piano. The construction of a guitar also shows the mathematics of sound. The frets of a guitar serve to change the lengths of its strings. When a musician presses a finger against a string, the string is pressed against a fret near the finger. This action effectively cuts off the string at the fret, allowing only the string below the fret to sound. The musician makes different notes from the guitar strings by using different frets.

The top string of a guitar is the E-string. When it is open (not pressed by a finger), it sounds an E. As you run a finger down the neck of the guitar on the E-string, you will get successively higher notes of the scale at each fret—F, F#, G, G#, and so on. In other words, the frets on a guitar are arranged so that the wavelength of sound produced by a string pressed against one fret is about 1.0595 times the wavelength of sound produced at the very next fret (moving toward the sound hole). Can you write an equation that describes this situation? Use the diagram on page 1 and let n stand for the number assigned to the frets, w_0 represent the wavelength of an open E, and w_n stand for the wavelength of a string sounding from the nth fret. (The equation is $w_n = w_0(1.0595)^n$.)

Use your equation to find relative wavelengths for E-notes from $n = 0$ to $n = 60$. Graph those points on page 2. Also graph the B-notes and G-notes indicated. Sketch a smooth curve through the points you've graphed. What kind of curve do you get? (It should be an exponential curve.)

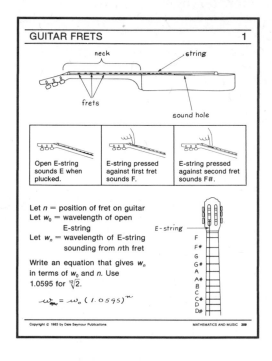

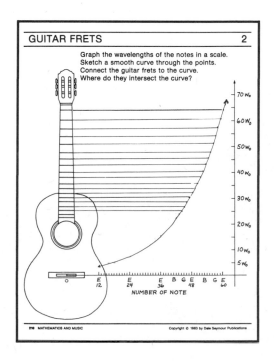

You may have noticed that the frets on a guitar are not spaced evenly. The spacing is related to the graph of $w_n = w_0(1.0595)^n$. Connect the guitar frets in the diagram on page 2 to the graph you drew. Explain the results. (The frets connect to the corresponding wavelengths of the notes they make.)

EXERCISES

1. The equation $f_n = (261.6)(1.0595)^{n-1}$ where n is an integer from 1 to 13 approximates the frequencies of notes in the middle octave of a piano. Approximate the frequencies to the nearest tenth (or get them from your work in the lesson on Piano Relationships). Graph the frequencies on a line.

2. The equation $w_n = w_0(1.0595)^n$ approximates the wavelengths of notes in the musical scale where w_0 stands for the wavelength of a high open E on the guitar and n stands for the number assigned to the frets of a guitar. Determine the relative wavelengths (values of $(1.0595)^n$) for the first 13 frets of a guitar. Graph the wavelengths on a line.

3. Compare the graph from Exercise 1 to the graph from Exercise 2.

4. In music, frequencies of notes and wavelengths of notes double over a full twelve-note octave. Within the octave there is an increase of about 5.95% for each successive note. In business, an investment will approximately double over a period of 12 years when the investment increases 5.95% each year. Write a formula to describe the business investment. Use n for number of years, P for the original amount of money invested, and A for the amount of money after n years. Compare your formula to the formula describing wavelengths of notes in a musical scale.

ANSWERS

1. See student work. The frequencies approximate a logarithmic scale.

2. Correct to two decimal places the wavelengths are $1.06w_0$, $1.12w_0$, $1.19w_0$, $1.26w_0$, $1.34w_0$, $1.41w_0$, $1.50w_0$, $1.59w_0$, $1.68w_0$, $1.78w_0$, $1.89w_0$, and $2.00w_0$. Student graphs should approximate a logarithmic scale.

3. The graphs are alike.

4. In business $A = P(1.0595)^n$. In music, $w_n = w_0(1.0595)^n$. Although the formulas describe totally different situations, they are alike; only the letters used to represent the values are different.

5. In general, for an investment with interest compounded annually, C dollars will approximately double in n years at p percent whenever p is an integral divisor of 72. This situation is often called the *Rule of 72*. Complete the following table based on the Rule of 72.

C dollars	p percent	n years	return on investment
100	6	12	200
100	8		200
100	3		200

6. According to the Rule of 72, at what rate of interest will an investment of $1000 double (approximately) when the interest is compounded annually and the money is invested for two years?

5. The values are 9 and 24.

6. 36%; in fact, at 36% the return on investment will be $849.60; at 40% the return on investment will be $1960 and at 41.42% the return will be $1999.9616. (These results can be obtained by solving $2000 = 1000(1 + r)^2$ for r.)

THE PIPE ORGAN

Vibrating strings create sound. Vibrating columns of air create sound too. Short columns of air produce high sounds. Long columns of air produce low sounds. This principle is used in a pipe organ. The length of organ pipes varies inversely as the pitch (measured by frequency in cycles per second). If you double the length of pipe, you halve the pitch. If you halve the length of pipe, you double the pitch. The exact relationship between length, L, and pitch, p, can be described by the following equation.

$$L = \frac{512}{p}$$

Using this equation, you can graph the relationship between length of pipe and pitch. Complete the table of values on page 1 and sketch the graph in the first quadrant. On what kind of curve do the points lie? (It's a hyperbola.)

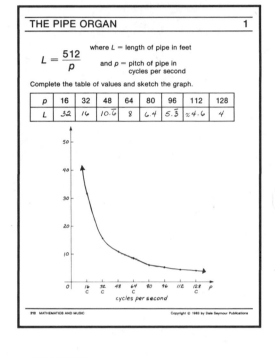

THE PIPE ORGAN   1

$L = \dfrac{512}{p}$ where L = length of pipe in feet
and p = pitch of pipe in cycles per second

Complete the table of values and sketch the graph.

p	16	32	48	64	80	96	112	128
L	32	16	10.6̄	8	6.4	5.3̄	≈4.6	4

cycles per second

213 MATHEMATICS AND MUSIC   Copyright © 1983 by Dale Seymour Publications

EXERCISES

1. The graph of $L = 512/p$ is a hyperbola. In what two quadrants of the coordinate plane does the graph lie?

2. Why does the graph of $L = 512/p$ representing the relationship between the length of organ pipes to their pitch lie only in the first quadrant of the coordinate plane?

3. What is the pitch (in cycles per second) of an organ pipe that is 10.7 feet long?

4. What length of organ pipe will give a pitch of 96 cycles per second?

ANSWERS

1. First and third

2. Measurements are positive numbers.

3. About 48 cycles per second

4. 5-1/3 feet

You have seen that the powers of $\sqrt[12]{2}$ (about 1.0595) are intimately related to the notes of the musical scale (by frequency and wavelength). You have used powers of $\sqrt[12]{2}$ to obtain some very interesting and useful graphs. There is another way to graph these values that gives a beautiful and fascinating picture.

The graph you will draw can be easily described using *polar coordinates*. (Recall that polar coordinates are of the form (r, θ) where r represents the distance, either positive or negative, of the point from the pole and θ represents the measure of the angle of rotation or angle formed between the polar axis and a ray connecting the point to the pole.) The polar coordinates of each point on your musical graph take the form $((\sqrt[12]{2})^n, n(15°))$. Complete the table of values on page 1 for n from 0 to 20. Graph the points you obtain and connect them with a smooth curve. The result should be a musical spiral. (In case you are wondering, the choice of 15° increments is arbitrary; 10° increments will generate more points; 30° increments will generate fewer points.)

If you look carefully at your graph, you will discover that each radius intersects the graph at an angle of approximately 77.5°. The radii intersect the graph at equal angles. For this reason, the spiral you have sketched is often called an *equiangular spiral*. Another name, and probably the most common, is *logarithmic spiral*. The curve has this name because the angles the radii form with the polar axis are proportional to the logarithms of the lengths of the corresponding radii. For example, when the angle is $2 \times 15°$, the corresponding log is $2 \times \log(1.0595)$ where 1.0595 approximates $\sqrt[12]{2}$; when the angle is $3 \times 15°$, the corresponding log is $3 \times \log(1.0595)$; and so on.

Logarithmic spirals are not only musical, they're golden. You can use a Golden Rectangle to generate a logarithmic spiral. You simply follow these steps.

1. Draw a Golden Rectangle *ABCD*.
2. Draw diagonal $\overline{DB}$.

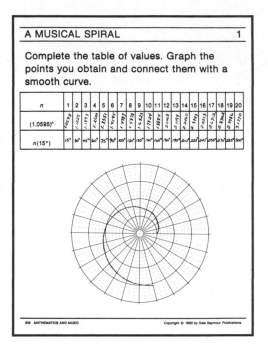

A MUSICAL SPIRAL   1

Complete the table of values. Graph the points you obtain and connect them with a smooth curve.

n	1	2	3	4	5	6	7	8	9	10	11	12	13	14	15	16	17	18	19	20
$(1.0595)^n$																				
$n(15°)$	15°	30°	45°	60°	75°	90°	105°	120°	135°	150°	165°	180°	195°	210°	225°	240°	255°	270°	285°	300°

3. Locate E on $\overline{DC}$ and F on $\overline{AB}$ so that figure $AFED$ is a square. Notice that the small rectangle formed, rectangle $FBCE$, is a Golden Rectangle. (Can you verify this fact?)

4. Draw diagonal $\overline{CF}$.

5. Locate G on $\overline{CB}$ and H on $\overline{EF}$ so that figure $ECGH$ is a square. This creates a new, even smaller Golden Rectangle, rectangle $HFBG$.

6. Continue creating squares and Golden Rectangles as long as space allows.

7. Starting with point F, draw an arc of a circle with radius equal to the side of square $AFED$. The arc should start at A and end at E. Continue the process for H, J, and so on.

Page 2 shows rectangle $ABCD$ with $\overline{EF}$ drawn. Complete steps 4–7 for this diagram. The completed spiral is an approximation of a logarithmic spiral. (The real logarithmic spiral is not made up of circular arcs, but its shape is very close to what you've drawn. The spiral does pass through A, E, B, and so on. In other words, the locus of these points *is* a logarithmic spiral when rectangle $ABCD$ is a Golden Rectangle.)

Another way to approximate a logarithmic spiral is by using a sequence of numbers closely related to the Golden Ratio—the Fibonacci numbers. Consider the ratio of two consecutive Fibonacci numbers such as 89/55. You will obtain a decimal approximation for the Golden Ratio that is correct to three decimal places—1.618. A rectangle measuring 89 units by 55 units is, then, a fair approximation to a Golden Rectangle. If, within this rectangle, you cut off a 55-unit-by-55-unit square, you obtain a smaller rectangle that is 55 units by 34 units. The value of 55/34 rounded to three decimal places is 1.618, so this 55 by 34 rectangle again is a fair approximation to the Golden Rectangle.

Cutting off a 34-unit-by-34-unit square of the small Golden Rectangle leaves a second Golden Rectangle measuring 34 units by 21 units, for which the ratio 34/21 rounded to three decimal places is 1.619—a fair approximation to a third Golden Rectangle. Repeating this process about six more times gives you five more Golden Rectangle approximations. Drawing quarter-circle arcs within each square will give you a fair approximation to a logarithmic spiral. Page 3 shows the first few rectangles. Complete the picture to see the approximation.

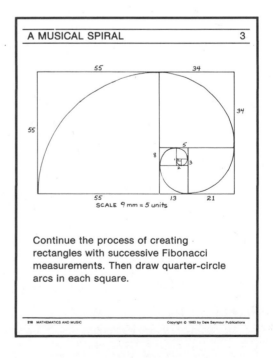

A MUSICAL SPIRAL · 3

SCALE 9 mm = 5 units

Continue the process of creating rectangles with successive Fibonacci measurements. Then draw quarter-circle arcs in each square.

Lest you think the possibilities for constructing logarithmic spirals are exhausted, take a look at the diagram on page 4. The points A, B, C, D, E, and so on are the locus of a logarithmic spiral. The isosceles triangles have been constructed so that $GF = 1\phi$, $FE = 1\phi + 1$, $ED = 2\phi + 1$, $DC = 3\phi + 2$, $CB = 5\phi + 3$, and so on. What kind of triangles have been constructed? (They are Golden Triangles because, for example, $DC/ED = (3\phi + 2)/(2\phi + 1) = \phi^4/\phi^3 = \phi$.)

Because logarithmic spirals maintain their shapes as they grow in size, they are a favorite of nature. Some spiders spin logarithmic spirals. The spider builds an outer frame and a center region. Then she spins radii from the center to the frame. Finally, she spirals from the center, working her way out to the frame. The shell of a chambered nautilus is a beautiful and extremely popular logarithmic spiral. The cutaway view on page 5 shows the spiraling shell and the chambers in which the nautilus lives. At any time in the creature's life, it inhabits only the largest outermost chamber. As it grows, it creates ever larger chambers, abandoning its old "room" for its newer bigger one. Throughout the natural world—from galaxies to ocean waves and from pine cones to canaries—there are logarithmic spirals to be seen; you only need to look.

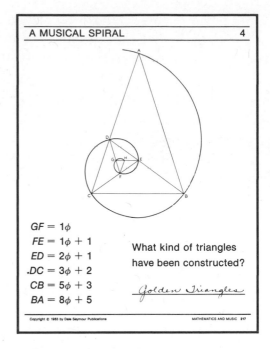

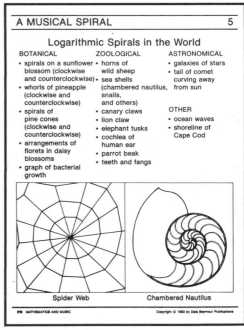

EXERCISES

1. Find the polar coordinates of a point on the musical spiral for $n = 40$.

2. Find the polar coordinates of a point on the musical spiral for $n = 100$.

ANSWERS

1. $(10.08, 600°)$, which is the same as $(10.09, 240°)$.

2. $(322.5, 1500°)$, which is the same as $(322.5, 60°)$.

3. Theoretically, does the musical spiral ever end?

4. How can you extend the logarithmic spiral drawn in the Golden Rectangles to make it larger?

5. How can you extend the logarithmic spiral drawn in the Fibonacci rectangles to make it larger?

6. Can you continue the process described for Exercise 5 indefinitely?

7. If you continue the process described for Exercise 5, where would the very next square be added to the figure?

8. The general equation for a point (r, θ) in polar coordinates on a logarithmic spiral is $r = e^{\theta \cot K}$ where K is the measure of the angle that any radius makes with the spiral curve. Approximate K for the point $(1.0595, \pi/12)$.

3. no

4. Construct a square on the bottom of rectangle $ABCD$ with one side $\overline{AB}$. Then, draw a quarter-circle from B to the opposite vertex.

5. Add an 89-by-89 square to the bottom of the largest rectangle.

6. Yes, assuming you have an unlimited size to your sheet of paper.

7. The square should be added along the rightmost edge of the figure. The sides of the square are 144 units long.

8. 77.5°

COXETER'S GOLDEN SEQUENCE

In 1968, H. S. M. Coxeter, a well-known geometer, published a remarkable result. He was interested in a sequence of circles constructed so that each four consecutive circles are tangent to one another. What is the radius of each circle and how are they positioned? Coxeter showed that the radii of the circles in the completed construction form a geometric sequence; each radius after the first can be found by multiplying the value of its predecessor by a common ratio. Coxeter showed that only one particular sequence of radii fits the conditions of the problem. He gave the value of the sequence's common ratio. To me, Coxeter's result is almost unbelievable. The common ratio is $\phi + \sqrt{\phi}$. It's the sum of the Golden Ratio and its square root!

The formula, then, for Coxeter's sequence of radii is

$$a_n = a_1(\phi + \sqrt{\phi})^{n-1}$$

where n is a positive
integer and a_n
stands for the nth radius.

Suppose you start with $a_1 = 0.1$ and use 1.618034 for ϕ. In this case, $a_n \approx (0.1)(2.890054)^{n-1}$. What are approximate values for the first six terms of the sequence? (They are 0.1, 0.289, 0.835, 2.414, 6.976, and 20.162 correct to three decimal places.) Can you draw Coxeter's Golden Sequence of tangent circles?

Actually constructing the sequence is not an easy task. Page 2 shows the completed construction for the first five circles and part of the sixth. The diagram shows you something else, too. The broken curve is a logarithmic (or equiangular) spiral. The sequence of contact points for the circles (points of tangency) lie on a logarithmic spiral!

COXETER'S GOLDEN SEQUENCE — 1

Let $a_1, a_2, a_3, \ldots$ be radii of tangent circles defined as follows:

1. Every 4 consecutive circles are mutually tangent.

2. The radius of each circle is obtained by multiplying the radius of the preceding circle by the sum of the golden ratio and its square root.

$$a_n = a_1(\phi + \sqrt{\phi})^{n-1}$$ where n is a positive integer and a_n stands for the nth radius

Using 1.6180340 for ϕ gives the following approximation for the general term.

$$a_n \approx a_1(2.890054)^{n-1}$$

Suppose $a_1 = 0.1$. Complete the following table.

term	a_1	a_2	a_3	a_4	a_5	a_6
approximate value	0.1	0.289	0.835	2.414	6.976	20.162

Can you draw Coxeter's golden sequence of tangent circles?

COXETER'S GOLDEN SEQUENCE — 2

Coxeter's Golden
Sequence of Tangent Circles

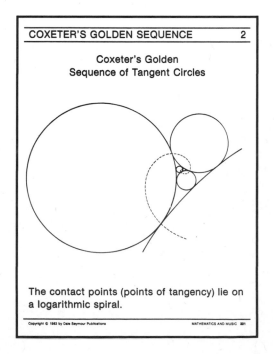

The contact points (points of tangency) lie on a logarithmic spiral.

EXERCISES

1. Letting a_1, a_2, a_3, ... stand for the radii of tangent circles in Coxeter's sequence, find values for a_6, a_7, and a_8 if $\phi = 1.618034$. Round your answers to the nearest three decimal places.

2. From the values you obtained in this lesson, calculate a_4/a_3.

3. Suppose you use 2.89 for the common ratio in Coxeter's sequence rather than 2.890054. In rounding to three decimal digits, which, if any, of the values for a_1, a_2, a_3, a_4, and a_5 differ?

ANSWERS

1. 20.162, 58.269, 168.400

2. 2.891018

3. The values for a_4 and a_5 will each differ by one-thousandth.

MASTERS

PROTECTING
A COMPUTER'S SECRETS

Number of Cities	Diagram	List of Routes		Number of Routes
2		AB BA		2 or 2!
3	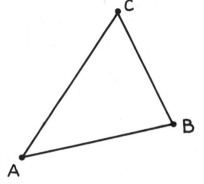	ABC ACB BAC BCA CAB CBA		6 or ___ !
4		ABCD ABDC ACBD ACDB ADBC ADCB	CABD CADB CBAD CBDA CDAB CDBA	——— or ___ !
		BACD BADC BCAD BCDA BDAC BDCA	DABC DACB DBAC DBCA DCAB DCBA	

Number of Cities	Diagram	List of Routes	Number of Routes
5	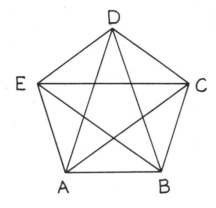	Sample: ADCEB	_____ or ___ !
10	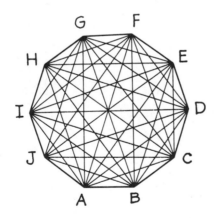	Sample: ABEDCFIGHJ	_____ or ___ !

How many seconds on a computer at
one million routes per second?

_____

For 18 cities there are ___! or about
_____ $\times$ 10$^-$ routes.

At one million routes per second, it
would take a computer about _____
years to test all of these routes.

For 60 cities, there are _____! routes.
At one million routes per second it
would take

    a. years.
    b. decades.
    c. centuries.
    d. billions of centuries.

How much time for 60 cities?

1. Find the number of routes.

   $60! \approx$ _____ $\times\ 10^{—}$

2. Find the number of seconds in a year.

$$\underline{\hspace{2cm}} \times \underline{\hspace{2cm}} \times \underline{\hspace{2cm}} = \underline{\hspace{2cm}}$$

seconds in one hour	hours in one day	days in one year	seconds in one year

3. Divide the number of routes by both the number of seconds in a year and the number of routes the computer can check in a second.

$$\cfrac{\text{routes}}{\cfrac{\text{seconds}}{\text{year}} \times \cfrac{\text{routes}}{\text{second}}}$$

= _____ years

= _____ centuries

= _____ vigintillion centuries

FINDING THE SHORTEST ROUTE

1. Find the number of routes that pass once and only once through each of 25 known cities.

2. Suppose a sales representative, Mr. Calling, knows the distance between each pair of the 25 cities described in Exercise 1. He wishes to determine the shortest route. Find the time needed for a computer to check each complete route if it can check one million routes per second.

3. Suppose that in one zone a telephone company has 50 pay phones from which coins must be collected periodically. The collection supervisor who picks up the coins wishes to find the shortest route to follow when visiting each pay phone on the route exactly once.

   a. Find the number of routes to be tested.

   b. Determine the computer time needed to check the length of each route. (Assume the computer can check one million routes per second.)

With one ball (◯) and two bins (⊔⊔) there are 2^1 or 2 ways to pack.

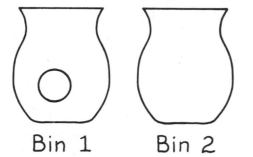

Bin 1   Bin 2

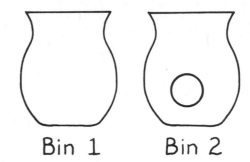

Bin 1   Bin 2

With two balls (◯ ●) and two bins (⊔⊔) there are 2^- or _____ ways to pack.

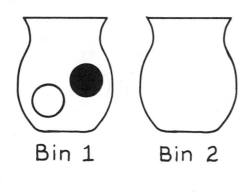

Bin 1   Bin 2

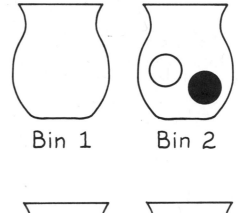

Bin 1   Bin 2

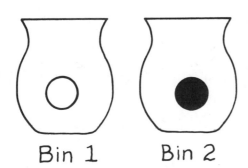

Bin 1   Bin 2

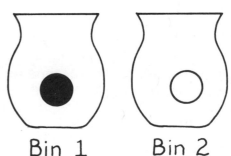

Bin 1   Bin 2

With three balls (○ ● ⊚)
and two bins (⛃⛃)
there are 2⁻ or ____ ways to pack.

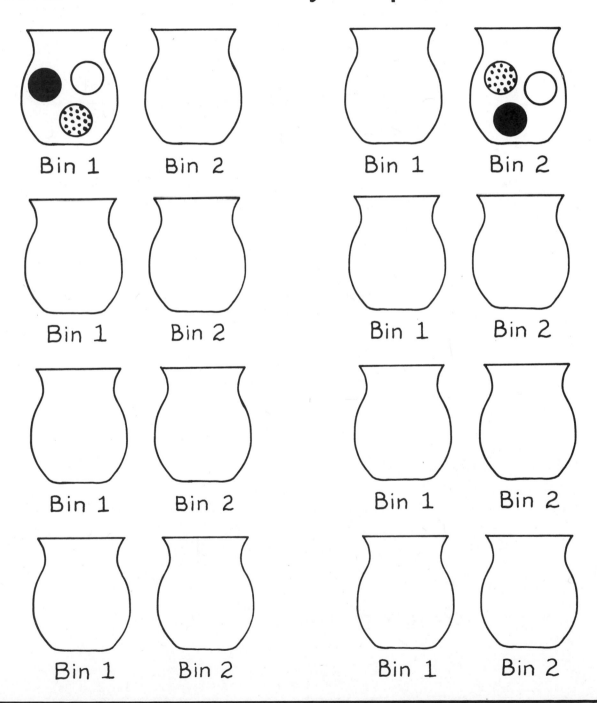

number of bins	number of balls	number of ways to pack	approximate computer time required
2	1	2	
2	2	2^2 or 4	
2	3	2^3 or 8	
2	4	2^4 or 16	
2	5	2^5 or 32	
⋮	⋮	⋮	
2	10	2^{10} or 1024	$\dfrac{1}{1000}$ second
⋮	⋮	⋮	
2	20	2^{20} or 1,048,576	1 second
⋮	⋮	⋮	
2	30	2^{30} or 1.0737×10^9	_____ minutes
⋮	⋮	⋮	
2	100	2^{100}	_____ centuries

The number of ways you can pack balls into two bins of indefinite size is a function of the number of balls you use. The function is an exponential function, $f(x) = 2^x$ where x represents the number of balls.

1. Draw a diagram showing all 16 different ways in which four Ping-Pong balls can be packed into two bins. (To check yourself as you work, it's helpful to know how many different ways n items can be chosen r at a time. A formula for finding this answer is as follows.)

$$_nC_r = \frac{n(n-1)(n-2)\cdots(n-r+1)}{r!}$$

2. Draw a diagram showing all 32 different ways in which five Ping-Pong balls can be packed into two bins.

3. Find the number of different ways 50 Ping-Pong balls can be packed into two bins. Write your answer in scientific notation.

4. Find the approximate computer time required to test all the packings of 50 Ping-Pong balls into two bins using a computer that tests one million packings per second.

x	1	2	3	4	5	6
f(x)	2	7	12	17	22	27

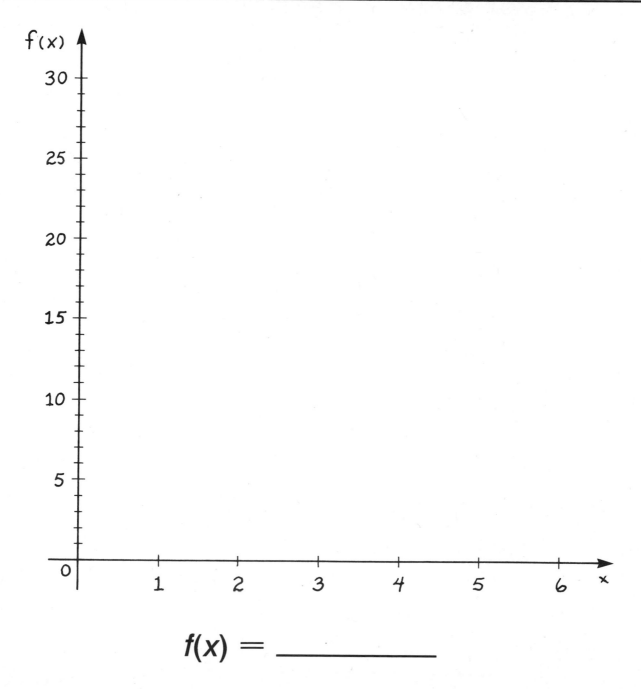

$$f(x) = \underline{\hspace{3cm}}$$

x	1	2	3	4	5	6
f(x)	2	7	12	17	22	27
g(x)	2	0	5	3	1	6

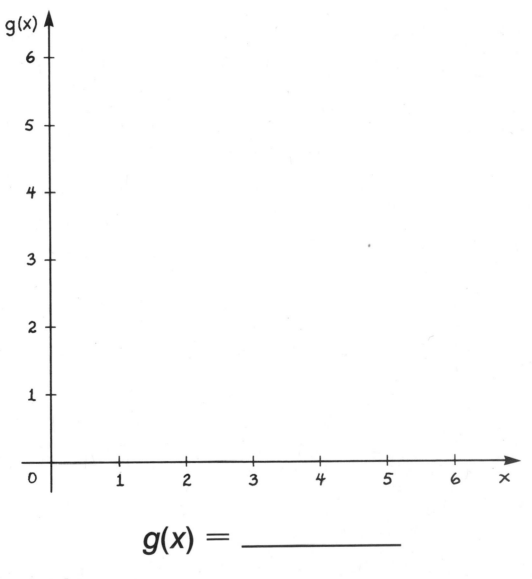

$$g(x) = \underline{\hspace{3cm}}$$

Try the rule:

1. Find an equation of the form $f(x) = mx + b$ for the linear function that describes the following transformation.

x	1	2	3	4	5	6
$f(x)$	7	10	13	16	19	22

2. Find an equation of the form $h(x) = ax^2 + b$ for the quadratic function that describes the following transformation.

x	1	2	3	4	5	6
$h(x)$	-4	2	12	26	44	66

3. Suppose $g(x) = f(x)$ mod 6 and $f(x) = 5x - 4$. Complete the following table.

x	1	2	3	4	5	6	7
$f(x)$	1	6	11		21		
$g(x)$	1	0		4			

4. Suppose $g(x) = f(x)$ mod 9 and $f(x) = 2x^3$. Complete the following table.

x	1	2	3	4	5	6
$f(x)$	2	16	54		250	
$g(x)$	2	7		2	7	

5. In 1982, several mathematicians devised a rapid solution to an NP problem. What was the problem and how long did it take to solve?

PACKING PROBLEMS

1
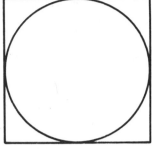
diameter = 1
density =

2
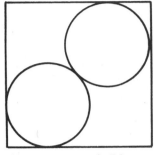
diameter = 0.58+
density =

3
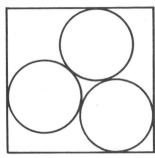
diameter = 0.5+
density =

4
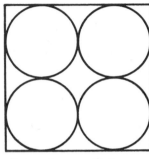
diameter = 0.5
density =

5
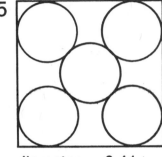
diameter = 0.41+
density =

6
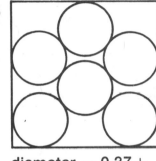
diameter = 0.37+
density =

7
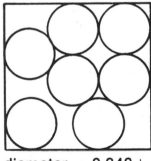
diameter = 0.348+
density =

8
diameter = 0.341+
density =

9
diameter = 0.33$\overline{3}$
density =

$$\text{density} = \frac{\text{sum of areas of circles}}{\text{area of square}}$$

$$= \frac{\text{number of circles} \times \text{area of one circle}}{\text{area of square}}$$

$$= \frac{n\pi r^2}{1^2}$$

FITTING CIRCLES INTO SQUARES

For each of the following exercises use 3.1415927 for π.

1. Find the density of the packing of 3 circles into a unit square (the packing as described in this lesson).

2. Find the density for the 4-circle packing.

3. Find the density for the 7-circle packing.

4. Find the density for the 8-circle packing.

5. Find the density for the 9-circle packing.

6. Goldberg had to decide how to best pack the circles into the square to come up with his figures. Find out how he suggested the circles should be packed. (The answer can be found in the January 1979 issue of *Scientific American* or in the January 1970 issue of *Mathematics Magazine*.)

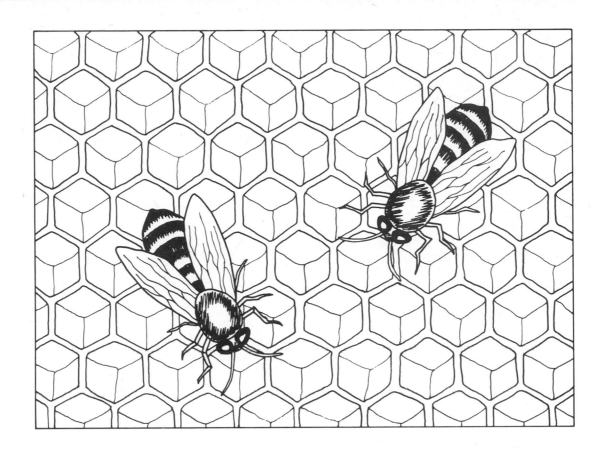

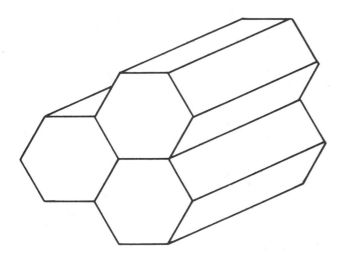

The top surfaces of regular hexagonal prisms form regular hexagons. Each angle of the hexagons measures _____°. As a result, the prisms fit together.

Tessellation	Density	Decimal Approximation

triangular

$$\frac{\text{area of circle}}{\text{area of triangle}}$$

or

$$\frac{\pi}{3\sqrt{3}}$$

0.6045998

square

$$\frac{\text{area of circle}}{\text{area of square}}$$

or

$$\frac{\pi}{4}$$

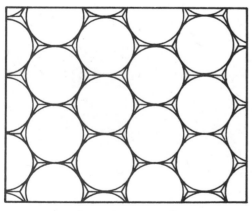

hexagonal

$$\frac{\text{area of circle}}{\text{area of hexagon}}$$

or

$$\frac{\pi}{2\sqrt{3}}$$

Which tessellation gives the densest packing?

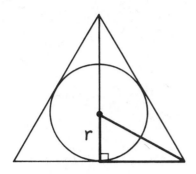

r = radius of inscribed circle

$$\text{density} = \frac{\text{area of inscribed circle}}{\text{area of triangle}}$$

1. Number of degrees in the angle of the small triangle.

   _____

2. Lengths of sides of the small triangle (in terms of r).

   _____

3. Length of one side of the equilateral triangle (in terms of r).

   _____

4. Formula for area of an equilateral triangle in terms of its sides.

   _____

Use the information above to find the density of a triangular tessellation.

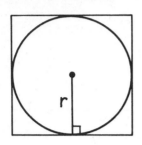

$r =$ radius of inscribed circle

density $= \dfrac{\text{area of inscribed circle}}{\text{area of square}}$

$=$

$=$

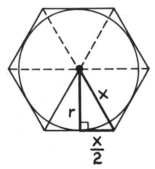

$r =$ radius of inscribed circle

density $= \dfrac{\text{area of inscribed circle}}{\text{area of hexagon}}$

$=$

$=$

Find the value
of x in terms
of r.

Use what you've
learned about
equilateral triangles.

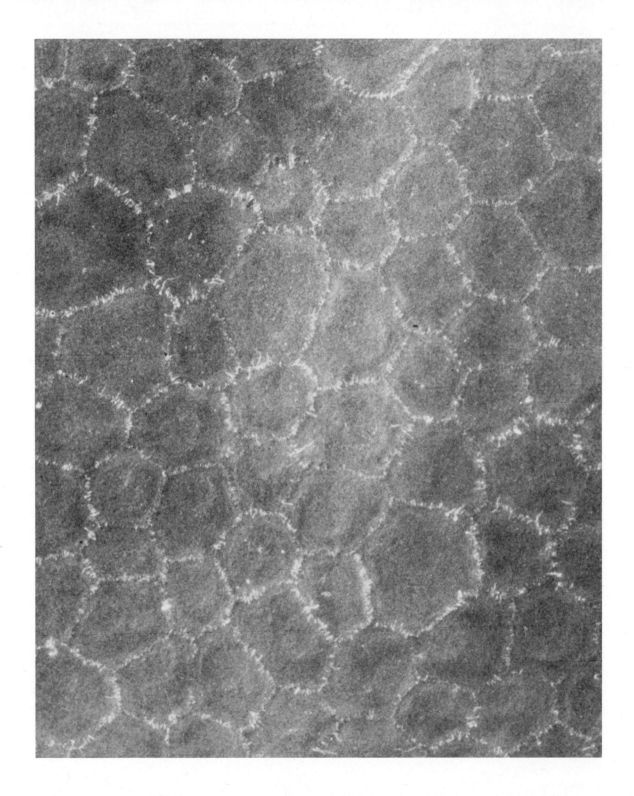

Endothelium Layer of Human Cornea

1. Explain why a regular pentagon cannot tessellate the plane.

2. Explain why a regular octagon cannot tessellate the plane.

3. Explain why the regular hexagon, the square, and the equilateral triangle are the only three regular polygons that can tessellate the plane.

4. There are many *nonregular* pentagons that will tessellate the plane. Give one example.

A VERY
SPECIAL NUMBER
CALLED *e*

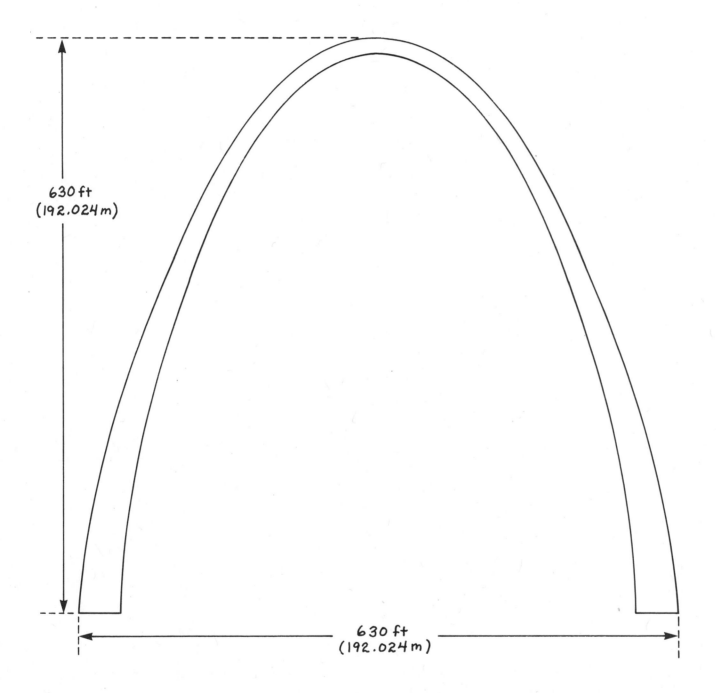

630 ft
(192.024 m)

630 ft
(192.024 m)

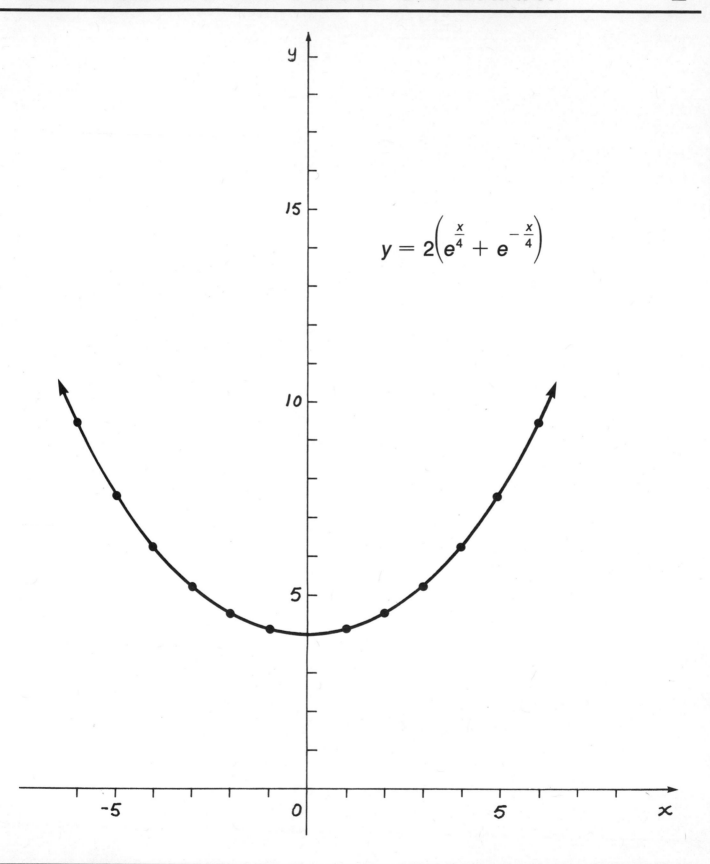

$$y = 2\left(e^{\frac{x}{4}} + e^{-\frac{x}{4}}\right)$$

THE FAMOUS
ST. LOUIS CATENARY   EXERCISES

1. Sketch the graph of a catenary curve for which $a = 1$. Find points for integer values of x from 0 to 4. Also find points for $x = 3.25$, $x = 3.5$, and $x = 3.75$. Use 2.718 for e.

2. Explain how the shapes of catenaries with $a = 2$ and $a = 1/2$ differ.

3. Explain how the graphs of catenaries with *positive* values of a differ from graphs of catenaries having *negative* values of a.

4. Give an equation for the catenary curve that has the same shape as $y = e^{x/2} + e^{-(x/2)}$ and opens in the same direction but has a y-intercept of 4.

$$y = ne^{kt} \text{ where } \begin{cases} y = \text{final amount} \\ n = \text{initial amount} \\ e \approx 2.72 \\ k = \text{constant} \\ t = \text{time elapsed} \end{cases}$$

Practical Applications:

- growth of bacteria
- cooling of coffee
- radioactive decay
- intensity of light passing through a medium
- atmospheric pressure at various heights above sea level
- electrical conductivity of glass at various temperatures
- decay of sound

For a certain strain of bacteria, the growth constant k is 0.867 when t is measured in hours. How long will it take for 20 bacteria to increase in number to 500 bacteria?

$$500 = 20\,(2.72)^{(0.867)t}$$
$$25 = 2.72^{(0.867)t}$$
$$\log 25 = \log 2.72^{(0.867)t}$$
$$= (0.867t)\,(\log 2.72)$$
$$t =$$

$$\approx$$

$$\approx$$

It will take about ___ hours.

1. Solve the bacterial growth problem for 40 bacteria growing to 600 bacteria.

2. Solve the bacterial growth problem for 15 bacteria growing to 400 bacteria.

3. Bacteria of a certain type can grow from 80 to 164 bacteria in 3 hours. Find the growth constant for this type of bacteria.

4. Tell whether the growth or decay constant is positive or negative for each of the following applications.
   a. decay of sound
   b. decomposition of radioactive substances
   c. intensity of light passing through a medium

5. In 10 years, the radioactive mass of a 200-gram sample decays to 100 grams. (This 10-year period is called the half-life of the radioactive material.) Write an equation that describes this situation and then find the decay constant for this radioactive substance.

$$x^y = y^x$$

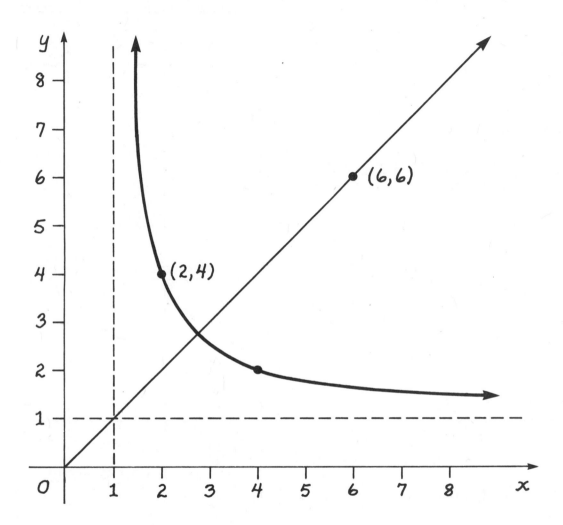

Does $2^4 = 4^2$? _____

Does $\left(\dfrac{9}{4}\right)^{\frac{27}{8}} = \left(\dfrac{27}{8}\right)^{\frac{9}{4}}$? _____

Does $\left(3\sqrt{3}\right)^{\sqrt{3}} = \left(\sqrt{3}\right)^{3\sqrt{3}}$? _____

These equations give positive *rational* solutions to $x^y = y^x$.

$$x = \left(1 + \frac{1}{n}\right)^{n+1}$$

$$y = \left(1 + \frac{1}{n}\right)^{n}$$

where $x > y$ and n is a positive integer.

n	x	y
1	4	2
2		
3		
4		

These equations give positive solutions to $x^y = y^x$, some of which are *irrational*.

$$x = k^{\frac{1}{(k-1)}}$$

$$y = k^{\frac{k}{(k-1)}}$$

where k is a positive real number and $k \neq 1$.

k	x	y
$\frac{4}{3}$		
2		
3		
4		
5		

1. Verify that $(-2, -4)$ and $(-4, -2)$ are solutions to the equation $x^y = y^x$.

2. Show that $(9/4, 27/8)$ is a solution to $x^y = y^x$.

3. Show that $(3\sqrt{3}, \sqrt{3})$ is a solution to $x^y = y^x$.

$$\pi = 3.141592653589793238\ldots$$
$$e = 2.718281828459045235\ldots$$

What are the sixteenth and seventeenth digits of π? _____

What are the sixteenth and seventeenth digits of e? _____

Compare $\pi^4 + \pi^5$ to e^6.

    What is $\pi^4 + \pi^5$ correct to 5 decimal places? _____

    What is e^6 correct to 5 decimal places? _____

Compare e^π to π^e.

    What is e^π correct to 6 decimal places? _____

    What is π^e correct to 6 decimal places? _____

x	$\dfrac{1}{x}$	$x^{\frac{1}{x}}$
1		
2		
2.1		
2.5		
2.9		
2.71		
2.718		
2.7182		
2.71828		
2.718281		
2.7182818		
2.8		
2.9		
3.0		

$x^{\frac{1}{x}}$ reaches its maximum value at $x = e$

$$e^{\frac{1}{e}} > \pi^{\frac{1}{\pi}} \qquad \underline{\hspace{3cm}}$$

$$\left(e^{\frac{1}{e}}\right)^{\pi e} > \left(\pi^{\frac{1}{\pi}}\right)^{\pi e} \qquad \underline{\hspace{3cm}}$$

$$e^{\pi} > \pi^{e} \qquad \underline{\hspace{3cm}}$$

1. Find the decimal value of 355/113 correct to 7 decimal places. Describe the relationship between this number and π.

2. Find the decimal value of 553/312 correct to 6 decimal places.

3. Find the value of $\sqrt{\pi}$ correct to 6 decimal places.

4. Use the results from Exercises 2 and 3 to describe the relationship between 553/312 and $\sqrt{\pi}$.

$$e^{ix} = \cos x + i \sin x$$

where x is a real number

Use Euler's formula to find the value of $e^{i\pi}$.

$$e^{i\pi} =$$

$$=$$

$$=$$

Euler's formula establishes the relationship among the four important numbers that are defined below. Which is which?

_____ $\lim\limits_{x \to \infty} \left(1 + \dfrac{1}{x}\right)^x$

_____ $\dfrac{\text{circumference}}{\text{diameter}}$

_____ additive inverse of the multiplicative identity for real numbers

_____ a solution to $x^2 + 1 = 0$

Approximate each partial sum. Use 3.14 for π.

Term of Sequence	Value
1	$1 + 0.00i$
$1 + \pi i$	$1 + 3.14i$
$1 + \pi i + \dfrac{(\pi i)^2}{2!}$	$-3.93 + 3.14i$
$1 + \pi i + \dfrac{(\pi i)^2}{2!} + \dfrac{(\pi i)^3}{3!}$	
$1 + \pi i + \dfrac{(\pi i)^2}{2!} + \dfrac{(\pi i)^3}{3!} + \dfrac{(\pi i)^4}{4!}$	
$1 + \pi i + \dfrac{(\pi i)^2}{2!} + \dfrac{(\pi i)^3}{3!} + \dfrac{(\pi i)^4}{4!} + \dfrac{(\pi i)^5}{5!}$	
$1 + \pi i + \dfrac{(\pi i)^2}{2!} + \ldots + \dfrac{(\pi i)^6}{6!}$	
$1 + \pi i + \dfrac{(\pi i)^2}{2!} + \ldots + \dfrac{(\pi i)^7}{7!}$	
$1 + \pi i + \dfrac{(\pi i)^2}{2!} + \ldots + \dfrac{(\pi i)^8}{8!}$	
$1 + \pi i + \dfrac{(\pi i)^2}{2!} + \ldots + \dfrac{(\pi i)^9}{9!}$	
$1 + \pi i + \dfrac{(\pi i)^2}{2!} + \ldots + \dfrac{(\pi i)^{10}}{10!}$	

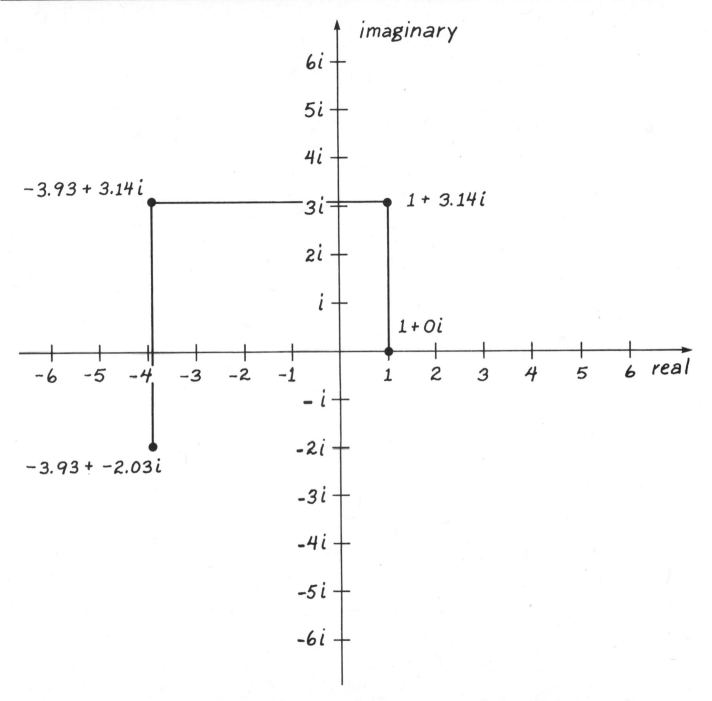

When connected in sequence, the terms form a counterclockwise spiral of line segments that approach _____ on the _____ axis.

EULER'S FORMULA EXERCISES

1. Express each of the following exponential expressions as a complex number in simplest form by using Euler's Formula.

   a. $e^{(\pi i)/2}$

   b. $e^{2\pi i}$

   c. $e^{(\pi/6)i}$

   d. $e^{(\pi/3)i}$

   e. $e^{(5/6)\pi i}$

   f. $e^{(4/3)\pi i}$

   g. $e^{(11/6)\pi i}$

   h. $e^{(9/4)\pi i}$

   i. $e^{3\pi i}$

2. Express each of the following sums as a power of e by using Euler's Formula.

   a. $\cos (2/3)\pi + i \sin(2/3)\pi$

   b. $\cos (\pi/4) + i \sin(\pi/4)$

   c. $\cos (3/2)\pi + i \sin(3/2)\pi$

   d. $\cos 4\pi + i \sin 4\pi$

   e. $(\sqrt{3}/2) + i(1/2)$ (Use the least value of x.)

$$i^{\,i} = e^{-\left(\frac{\pi}{2}\right)} \approx 0.2078795763\ldots$$

k	$i^{\,i}$ or $e^{-\left(\frac{\pi}{2}\right)+2k\pi}$ value	decimal approximation	$\frac{1}{i^{\,i}}$ or ____ value	decimal approximation
0	$e^{-\left(\frac{\pi}{2}\right)}$	0.2078798		
1	$e^{\frac{3\pi}{2}}$	111.31743		
−1	$e^{-\left(\frac{5\pi}{2}\right)}$	0.0003882		
2				
−2				
3				
−3				

1. Use Euler's Formula to show that $i^i = e^{-(\pi/2)}$.
   (HINT: Use $e^{-(\pi/2)} = e^{(\pi/2)i^2} = e^{[(\pi/2)i]i}$.)

2. Use Euler's Formula to show that $i^{1/i} = e^{\pi/2}$.
   (HINT: Use $e^{\pi/2} = e^{(\pi/2)(i/i)} = e^{[(\pi/2)i][1/i]}$.)

x	$\left(1 + \dfrac{1}{x}\right)^x$
1	2.0000000
2	2.2500000
3	2.3703704
4	2.4414063
5	2.4883200
⋮ 8	
⋮ 20	
80	
100	
1000	
50,000	
70,000	
80,000	

$$e = \lim_{x \to \infty} \left(1 + \frac{1}{x}\right)^x$$

1. Use a calculator to evaluate $[1 + 1/x]^x$ for $x = 90$.

2. Evaluate $[1 + (1/x)]^x$ for $x = 500$.

3. Evaluate $[1 + (1/x)]^x$ for $x = 25,000$.

4. Evaluate $[1 + (1/x)]^x$ for $x = 60,000$.

5. Evaluate $[1 + (1/x)]^x$ for $x = 75,000$.

6. Evaluate $[1 + (1/x)]^x$ for $x = 79,000$.

7. How accurate an approximation for *e* do you obtain if you use 10,000 for x?

ALGEBRAIC FUNCTIONS

Graph the data.

weekly paycheck in dollars	$21	$37	$13	$53
number of items sold in week	2	4	1	6

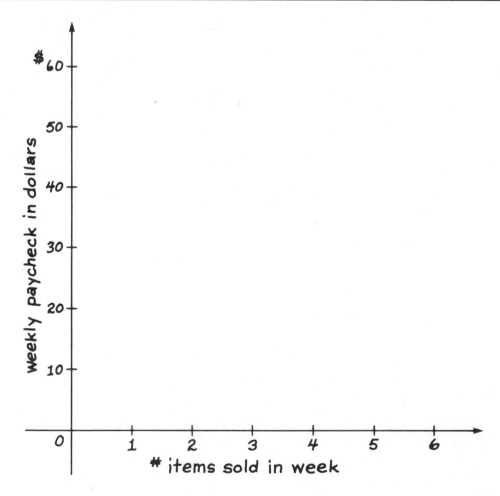

What equation best describes the data?

Graph the data.

math grades	95	51	49	27	42	52	67	48	46	69	82
science grades	88	70	65	50	60	80	68	49	40	75	81

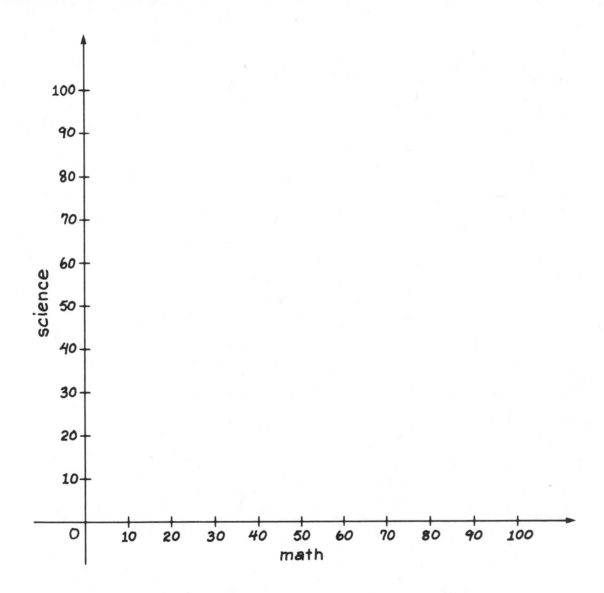

What equation best describes the data?

1. Graph the following set of data. Determine a linear equation that approximates the data.

x	33	45	46	20	40	30	38	22	52	44	55
y	4	7	8	1	6	3	5	2	9	6	10

2. Graph the following set of data. Determine a linear equation that approximates the data.

x	1	3	5	8	9	10	12	15
y	1.0	2.0	3.0	4.5	5.0	5.5	6.5	8.0

Find a graph in the January, 1980 *Scientific American* that shows a parabolic function.

Find a graph in the January, 1979 *Scientific American* that shows addition of functions.

Find a graph in the November, 1979 *Scientific American* that shows addition of functions.

Find a graph in the October, 1979 *Scientific American* that shows a logarithmic function.

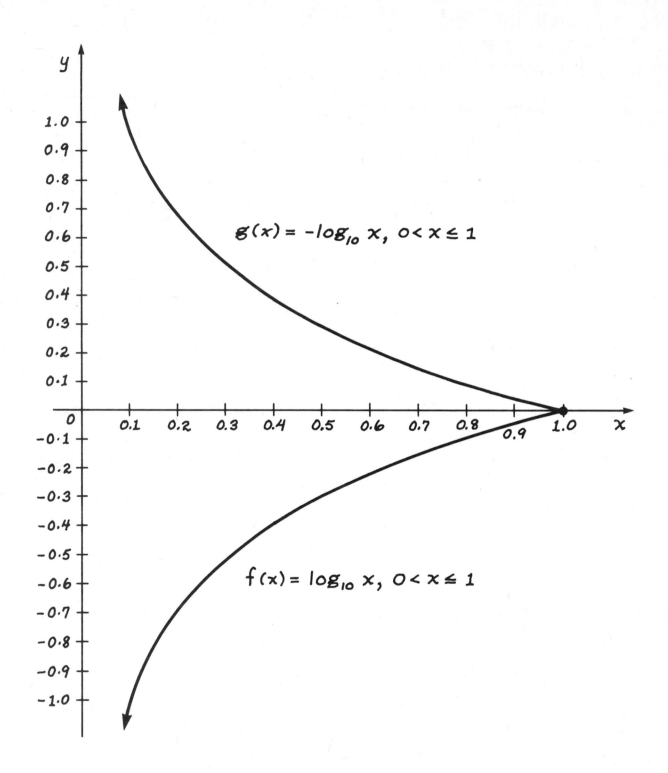

$g(x) = -\log_{10} x, \; 0 < x \leq 1$

$f(x) = \log_{10} x, \; 0 < x \leq 1$

ALGEBRAIC GRAPHS EXERCISES

1. For each parabola whose equation is given, find the maximum value, the maximum point, and name the axis of symmetry.
   a. $y = -2(x - 3)^2 + 4$
   b. $y = -(x + 5)^2 + 3$
   c. $y = -(5/4)(x - 7)^2 + (2/3)$
   d. $y = -(1/2)(x + 4)^2 - 3$

2. How would you describe the graph of the quadratic function whose equation is $y = 2(x - 4)^2 + 5$?

3. Graph the following functions on the same coordinate axes. Give the equation for the graph of *c*.
   a. $f(x) = (1/2)x + 1$
   b. $g(x) = (x - 2)^2$
   c. $(f + g)(x)$

4. Graph the following functions on the same coordinate axes. Explain how they are related.
   a. $f(x) = (1/3)x + 2$
   b. $-f(x)$

5. Graph the following functions on the same coordinate axes. Explain how they are related.
   a. $g(x) = (1/2)(x - 2)^2 + 1$
   b. $-g(x)$

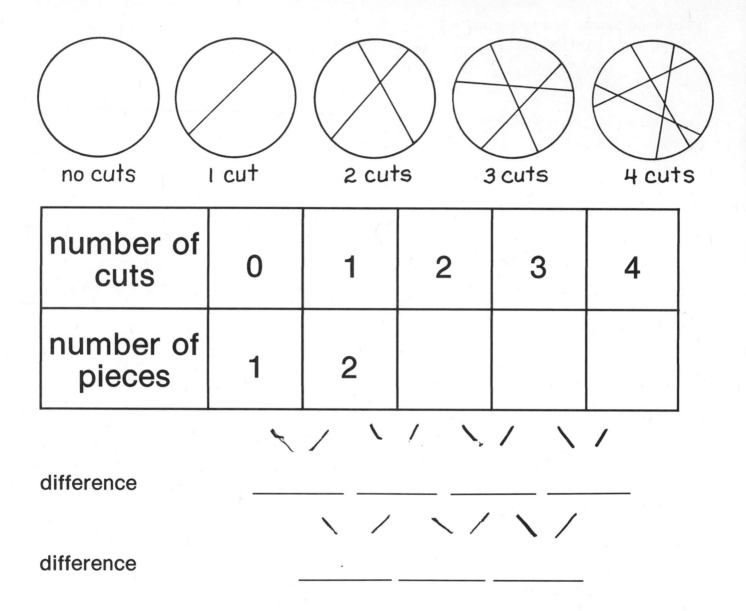

no cuts   I cut   2 cuts   3 cuts   4 cuts

number of cuts	0	1	2	3	4
number of pieces	1	2			

difference _____ _____ _____ _____

difference _____ _____ _____

What is the maximum number of pizza pieces you can obtain by making 5 cuts? _____

$$f(x) = ax^2 + bx + c$$

x	0	1	2	3	4
y	c				

difference _____ _____ _____ _____

difference _____ _____ _____

If the first numbers in each row are 1, 1, and 1, respectively, what is the quadratic relationship? _____

If the first numbers in each row are −2, 3, and 4, respectively, what is the quadratic relationship? _____

A QUADRATIC PIZZA FUNCTION

Ask a friend to *think* of a quadratic equation in the form $y = ax^2 + bx + c$. (Your friend should decide on particular values for a, b, and c.) Then ask your friend to tell you the values for y when x is 0, 1, 2, 3, and 4. Your friend should give you the values in order. Write the values on a piece of paper or a chalkboard for your friend (and audience) to see. Find the differences as you did in this lesson. From the differences, determine the original quadratic equation. Any friend who doesn't know the trick will be amazed.

1. What is the quadratic equation if your friend's values are −4, 1, 12, 29, and 52?

2. What is the quadratic equation if your friend's values are −3, 4, 15, 30, and 49?

3. What is the quadratic equation if your friend's values are 8, 7, 16, 35, and 64?

$$f(x) = ax^3 + bx^2 + cx + d$$

x	0	1	2	3	4
f(x)					

difference _____

difference _____

difference _____

Find the equation for the function given in this table.

x	0	1	2	3	4	5
f(x)	13	17	27	55	113	213

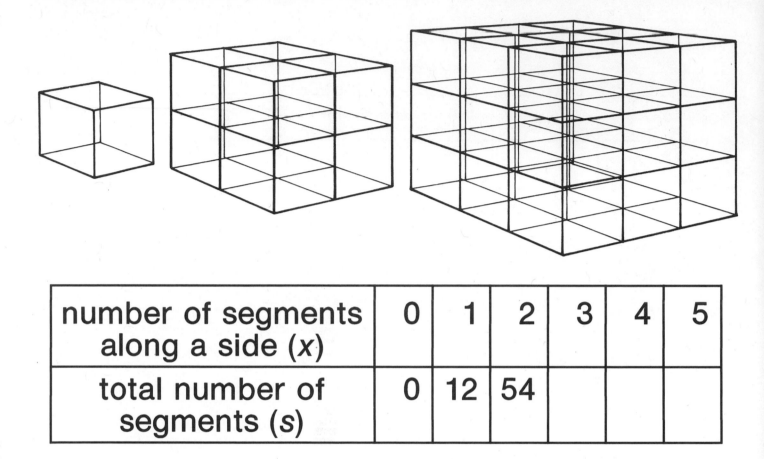

number of segments along a side (x)	0	1	2	3	4	5
total number of segments (s)	0	12	54			

What is the equation that gives the total number of individual segments, s, when the number of segments on a side, x, is known? _____

FINDING CUBICS

EXERCISES

1. Determine the cubic equation that could generate the following data.

x	0	1	2	3	4	5
y	−10	−22	−24	−10	26	90

2. Determine the cubic equation that could generate the following data.

x	0	1	2	3	4	5
y	14	3	−6	5	54	159

3. Find the number of individual segments within a cubic structure that has 6 segments on a side.

4. Find the number of individual segments within a cubic structure that has 7 segments on a side.

5. Find the number of individual segments within a cubic structure that has 8 segments on a side.

6. Find the number of individual segments within a cubic structure that has 9 segments on a side.

A group consists of a set of elements and a rule of combination, $*$, (binary operation) defined for pairs of elements for which the following properties are satisfied.

- **CLOSURE**
  The combination of any two elements of the set also belongs to the set.

- **ASSOCIATIVITY**
  When combining elements of the set, they can be grouped in any way.

  e.g. $a * (b * c) = (a * b) * c$

- **IDENTITY ELEMENT**
  There is an element of the set that when combined with any other element of the set leaves that element unchanged.

- **INVERSES**
  For any element of the set, there is another element that combines with it to produce the identity.

The integers under addition form a group.
What is the identity element?

How do you find the inverse of an element in that group?

A SPECIAL GROUP OF FUNCTIONS 2

The set of functions $\{f_1, f_2, f_3, f_4, f_5, f_6\}$ forms a group under composition of functions where the functions are defined as follows.

$$f_1(x) = x \qquad\qquad f_4(x) = \frac{x-1}{x}$$

$$f_2(x) = \frac{1}{x} \qquad\qquad f_5(x) = \frac{x}{x-1}$$

$$f_3(x) = 1 - x \qquad\qquad f_6(x) = \frac{1}{1-x}$$

Complete the operations table for this group.

*	f_1	f_2	f_3	f_4	f_5	f_6
f_1						
f_2			f_6			
f_3		f_4				
f_4						
f_5						
f_6				f_1		

What is the identity element? _____

What is the inverse of f_5? _____

A SPECIAL GROUP OF FUNCTIONS 3

The set of functions $\{f_1, f_4, f_6\}$ forms a subgroup.

Complete the operations table.

*	f_1	f_4	f_6
f_1			
f_4			
f_6			

What is the identity element? ____________

Name the inverse for each function.

inverse of f_1 ____________

inverse of f_4 ____________

inverse of f_6 ____________

Graph f_1, f_4, and f_6.

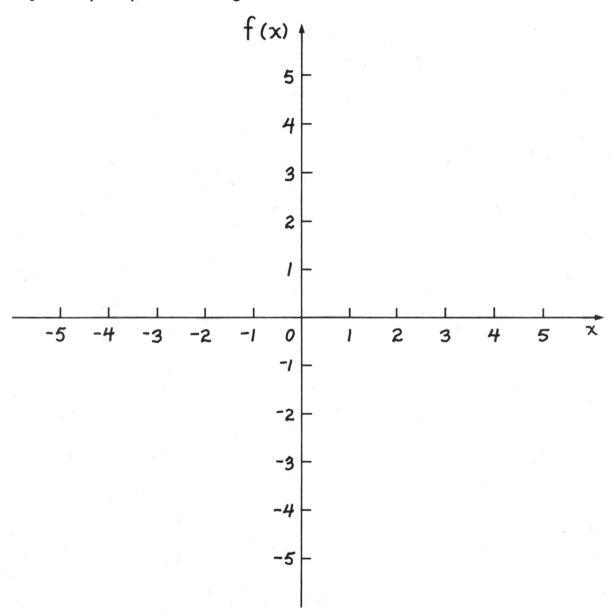

Describe the inverse relationship for this group in geometric terms.

1. Which of the following algebraic systems are groups?

   a. whole numbers under addition

   b. rational numbers under addition

   c. integers under multiplication

   d. rational numbers under multiplication

2. Show that $f_4 \cdot (f_6 \cdot f_4) = (f_4 \cdot f_6) \cdot f_4$.

3. Does the group of functions $f_1, f_2, f_3, f_4, f_5, f_6$ have any subgroups other than the one mentioned in this lesson? If so, name one.

THE GOLDEN RATIO

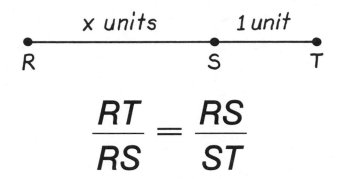

$$\frac{RT}{RS} = \frac{RS}{ST}$$

Substitute values.

Multiply means and extremes.

Solve the quadratic.

The length of $\overline{RS}$ is _____.

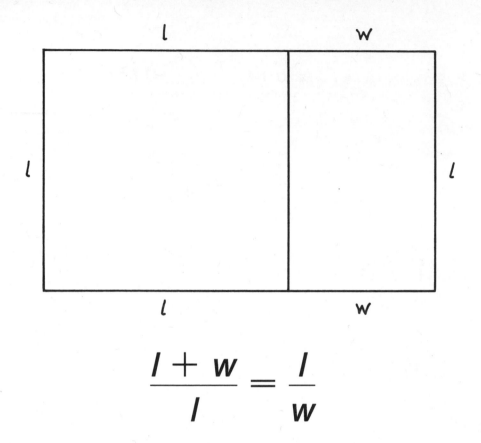

$$\frac{l + w}{l} = \frac{l}{w}$$

Solve for l.

WHAT IS THE GOLDEN RATIO?   EXERCISES

The equation $x^2 - x - 1 = 0$ is called the Golden Quadratic Equation.
We use the symbols ϕ and ϕ' to represent the solutions to the equation.

$$\phi = \frac{1 + \sqrt{5}}{2} \quad \text{and} \quad \phi' = \frac{1 - \sqrt{5}}{2}$$

1. Find $\phi + \phi'$.

2. Find $\phi \cdot \phi'$.

3. Show that $\phi - 1 = 1/\phi$.

4. Find a decimal approximation for the Golden Ratio correct to 6 decimal places.

5. Suppose a point C divides a line segment $\overline{AB}$ into two segments, $\overline{AC}$ and $\overline{CB}$, so that $AC/CB = AB/AC$. Show that each ratio in the proportion is a Golden Ratio.

$$1 + \cfrac{1}{1 + \cfrac{1}{1 + \cfrac{1}{1 + \cfrac{1}{1 + \ldots}}}}$$

Term	Decimal Value
1st: 1	1
2nd: $1 + \dfrac{1}{1}$	2
3rd: $1 + \cfrac{1}{1 + \dfrac{1}{1}}$	1.50
4th: $1 + \cfrac{1}{1 + \cfrac{1}{1 + \dfrac{1}{1}}}$	$1.\overline{6}$
5th:	
6th:	
7th:	

Let $x = 1 + \cfrac{1}{1 + \cfrac{1}{1 + \cfrac{1}{1 + \cfrac{1}{1 + \ldots}}}}$

$x \curvearrowright$

So $x = 1 + \dfrac{1}{x}$.

Rewrite the equation in quadratic form and solve for x.

CONTINUED FRACTIONS   EXERCISES

1. Find the rational number in fraction form that equals the following fraction.

$$1 + \cfrac{1}{2 + \cfrac{1}{3 + \cfrac{1}{4 + \cfrac{1}{5}}}}$$

2. Find the rational number in fraction form that equals the following fraction.

$$2 + \cfrac{1}{2 + \cfrac{1}{2 + \cfrac{1}{2 + \cfrac{1}{2}}}}$$

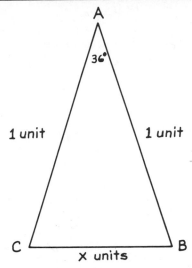

Use the Law of Cosines to find x.

$$BC^2 = AC^2 + AB^2 - 2(AB)(AC) \text{ Cos } A$$

$$x^2 = (\underline{\hspace{1cm}})^2 + (\underline{\hspace{1cm}})^2 - 2\,(\underline{\hspace{1cm}})(\underline{\hspace{1cm}}) \text{ Cos } \underline{\hspace{1cm}}$$

$$=$$

$$=$$

Compare the value of x to the reciprocal of the Golden Ratio.

$$x =$$

$$\frac{1}{\dfrac{1 + \sqrt{5}}{2}} =$$

1. Find the length of the side opposite the 36° angle in an isosceles triangle for which the two sides including this angle have lengths of 10 units. First predict the answer, then find it.

2. Explain why all Golden Triangles are similar.

3. Prove that AC/CB is the Golden Ratio for $\triangle ABC$ of this lesson.

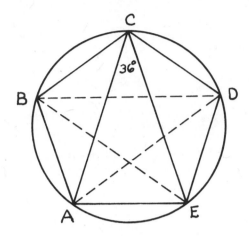

$\triangle ACE$ is a Golden Triangle

Name the four triangles that are congruent to $\triangle ACE$.

List all the Golden Ratios formed by these Golden Triangles.

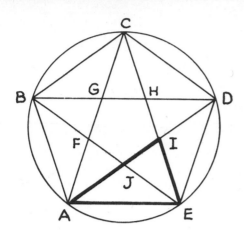

$\triangle AEI$ is a Golden Triangle. Name the triangles congruent to $\triangle AEI$. Count the Golden Ratios formed.

$\triangle CHG$ is a Golden Triangle. Name the triangles congruent to $\triangle CHG$. Count the Golden Ratios formed.

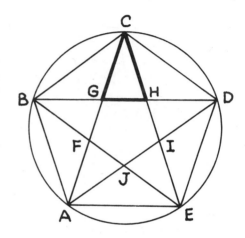

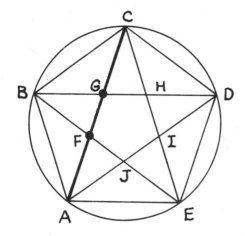

CH/HI, CI/IE, CE/CI, EI/IH, CI/CH, EH/EI are Golden Ratios. Which of these ratios have already been counted? How many new ratios are formed in all?

In all, _____ Golden Triangles and _____ Golden Ratios have been counted.

THE REGULAR PENTAGON   EXERCISES

The following chain of reasoning provides sufficient information for you to easily verify all of the statements in the discussion of the regular pentagon. Fill in the blanks with the correct information.

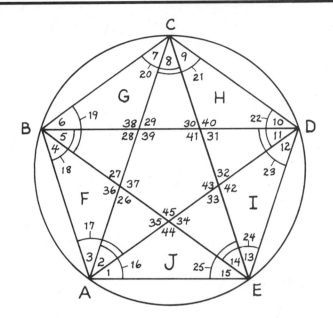

1. Since congruent chords of a circle cut off congruent arcs, each arc formed by the regular pentagon has a measure of 72. Since inscribed angles of a circle are measured by one-half of their intercepted arcs, Angles 1–15 each have a measure of _____ and Angles 16–25 each have a measure of _____.

2. Also, an angle formed by two intersecting chords is measured by one-half the sum of its intercepted arc and the intercepted arc of its vertical angle. Therefore, Angles 26–35 each have a measure of (1/2)(_____ + _____) or _____. Angles 36–45 each have a measure of (1/2)(_____ + _____) or _____.

3. By the definition of regular pentagon, segments $\overline{AB}$, $\overline{BC}$, _____, _____, and _____ are congruent.

4. It follows that by the angle-side-angle postulate for congruent triangles, triangles *AFB*, _____, _____, _____, and _____ are congruent.

5. Since corresponding sides of congruent triangles are congruent and two sides opposite two congruent angles in a triangle are congruent, $\overline{AF} = \overline{FB} = \overline{BG} = \overline{GC} = $ _____ = _____ = _____ = _____ = _____ = _____.

6. As a result, by the side-angle-side postulate for congruent triangles, triangles *JAF*, *FBG*, _____, _____, and _____ are congruent.

7. Corresponding parts of congruent triangles are congruent implies that segments $\overline{JF}$, $\overline{FG}$, _____, _____, and _____ are congruent.

8. Therefore, pentagon *FGHIJ* is equilateral and, since all of its angles have measures of _____, it is equiangular. Consequently, it is a regular pentagon.

9. If you draw in diagonals $\overline{GJ}$, _____, _____, _____, and _____, this new regular pentagon star will contain _____ Golden Triangles and _____ Golden Ratios.

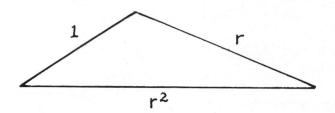

What values can *r* take?

Give reasons for each statement in CASE 1.

Then provide a similar argument for CASE 2.

CASE 1: $r \geq 1$

$$1 + r > r^2 \qquad \underline{\hspace{3cm}}$$

$$r^2 - r - 1 < 0 \qquad \underline{\hspace{3cm}}$$

$$\left(r - \frac{1 + \sqrt{5}}{2}\right)\left(r - \frac{1 - \sqrt{5}}{2}\right) < 0 \qquad \underline{\hspace{3cm}}$$

$$1 \leq r < \frac{1 + 5}{2} \qquad \underline{\hspace{3cm}}$$

CASE 2: $0 < r < 1$

Combine the results of the two cases.

$$\underline{\hspace{4cm}} < r < \underline{\hspace{4cm}}$$

1. How does the value of r for a triangle whose sides measure a, ar, and ar^2 where a is positive (any triangle whose sides have measures that form a geometric progression) compare to the value of r when $a = 1$?

2. Each threesome of numbers corresponds to the measures of the sides of a triangle. Tell whether or not the measures are in geometric progression, and if they are, give the ratio r.

sides of triangle	geometric?	r
1, 0.8, 0.64		
3, 2.4, 1.92		
4, 14, 16		
1/3, 1/4, 3/16		
1, 1, 1		
2, 1.4, 0.98		
7, 7, 10		
5, $5\sqrt{2}$, 10		

3. Find the solution set for $x^2 + 2x - 15 > 0$.

4. Find the solution set for $x^2 + 9x + 18 < 0$.

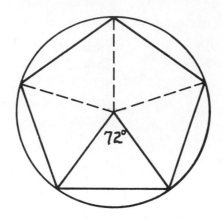

Compare Cos 72° to 1/2ϕ where ϕ is the Golden Ratio.

$$\frac{1}{2\phi} = \frac{1}{2\left(\dfrac{1 + \sqrt{5}}{2}\right)}$$

$$= \frac{1}{1 + \sqrt{5}}$$

$$= \frac{1}{1 + \sqrt{5}} \cdot \frac{1 - \sqrt{5}}{1 - \sqrt{5}}$$

$$= \frac{1 - \sqrt{5}}{-4}$$

$$\approx \underline{\hspace{3cm}}$$

$$\text{Cos } 72° \approx \underline{\hspace{3cm}}$$

THE CENTRAL ANGLE
OF A PENTAGON

1.–4. The discussion of the pentagon star inscribed in a circle gave many examples of ratios of line segments that are Golden Ratios. What about the angles of the figure in this lesson? (Connect all vertices to the center.) Are they in some way related to the Golden Ratio? Answer this question for yourself by completing the following table. (Note that the angle measuring 144° is supplementary to an angle measuring 36°, so it is included in this table for your interest.)

degree measure of angle	Cosine decimal form	Cosine radical form	Cosine ϕ form
72			
36			
108			
144			

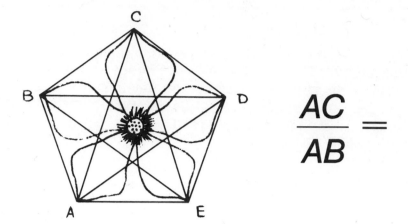

$$\frac{AC}{AB} =$$

hibiscus

wild geranium

spring beauty

wood sorrel

yellow violet

plumeria

saxifrage

wild rose

buttercup

polemonium

trailing arbutus

marigold

pipsissewas

bellflower

filaree

hoya plant

columbine

cinquefoil

chickweed

passion flower

Hoya Blossoms

Starfish

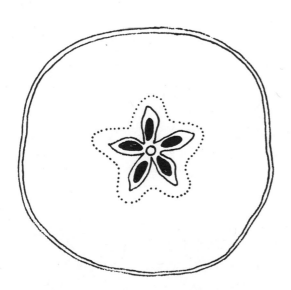

Cross Section of
Apple Seedbed

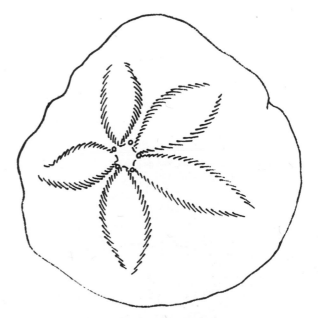

Sand Dollar

1. Find the names of ten flowers whose blossoms contain five petals and are *not* on the list in this lesson.

2. Name a plant that has either five petals on five additional petals or five petals on five sepals and does *not* appear on the list in this lesson.

3. Name two examples of pentagrams in nature that are *not* mentioned in this lesson.

THE 13TH CENTURY
STILL LIVES

1, 1, 2, 3, 5, 8, 13, 21, 34, _____, _____, . . .

Number of Petals	Name of Flower	Number of Petals	Name of Flower
2	Enchanter's Nightshade	21	Chicory Aster Helenium
3	Trilium Lily Iris	34	Plantain Ox-eye Daisy Pyrethrum
5	Wild Geranium Spring Beauty Yellow Violet	_____	Field Daisy Helenium Michaelmas Daisy
8	Lesser Celandine Sticktight Delphinium	_____	Michaelmas Daisy
13	Corn Marigold Mayweed Ragwort		

FIBONACCI, RABBITS, AND FLOWERS 2

SUNFLOWER SPIRALS

one way	8	21	34	55	89
other way	13	34	55	89	144

PINEAPPLE SPIRALS
5 and 8, 8 and 13

PINE CONE SPIRALS
5 and 8

How many spirals?

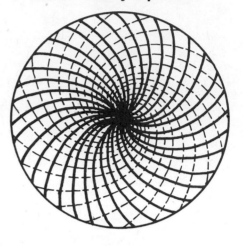

What is the ratio of turns to leaves on this stem?

LEAF ARRANGEMENTS ON TREE STEMS

Name of Tree	Turns : Leaves Ratio
basswood elm	1/2
hazel beech	1/3
apricot cherry oak	2/5
pear poplar	3/8
almond pussy willow	5/13

FIBONACCI, RABBITS, AND FLOWERS

EXERCISES

1. In Fibonacci's rabbit problem, how many pairs of rabbits will there be at the beginning of the seventh month? How many pairs will there be at the beginning of the twelfth month?

2. In Fibonacci's rabbit problem, how many pairs of adult rabbits (at least one month old) will there be at the beginning of the seventh month? How many baby rabbits (less than one month old) will there be?

3. Solve exercise 2 for the beginning of the twelfth month.

4. Fibonacci numbers have some remarkable properties. Find the missing numbers in this sequence of sums. Describe the pattern.

$$1^2 = 1 \times 1$$
$$1^2 + 1^2 = 1 \times 2$$
$$1^2 + 1^2 + 2^2 = 2 \times 3$$
$$1^2 + 1^2 + 2^2 + 3^2 = 3 \times \underline{\quad}$$
$$1^2 + 1^2 + 2^2 + 3^2 + 5^2 = \underline{\quad} \times \underline{\quad}$$
$$1^2 + 1^2 + 2^2 + 3^2 + 5^2 + 8^2 = \underline{\quad} \times \underline{\quad}$$
$$1^2 + 1^2 + 2^2 + 3^2 + 5^2 + 8^2 + 13^2 = \underline{\quad} \times \underline{\quad}$$

5. Find examples of at least 5 flowers named in this lesson. Find pictures from a book about flowers. Then try to find these flowers in gardens, along roadsides, or in parks. Count the petals. Are your results Fibonacci numbers?

6. Count the spirals, clockwise and counterclockwise, in a real pineapple. Are there 8 spirals one way and 13 the other?

7. Verify at least one *turns-to-leaves* ratio from the stem of a real tree that is listed in this lesson.

FIBONACCI NUMBERS & THE GOLDEN RATIO

1

n	$\dfrac{F_{n+1}}{F_n}$	Decimal Approximation
1	1/1	1.00000000
2	2/1	2.00000000
3	3/2	1.50000000
4	5/3	1.66666667
5	8/5	1.60000000
6	13/8	1.62500000
7	21/13	1.61538462
8	34/21	1.61904762
9	55/34	
10	89/55	
11		
12		
13		
14		
15		

$$\lim_{n\to\infty} \frac{F_n + 1}{F_n} = \underline{\hspace{3cm}}$$

$$\approx \underline{\hspace{3cm}}$$

THE THIRTEENTH CENTURY STILL LIVES   Copyright © 1983 by Dale Seymour Publications

$$\phi^1 \approx 1.6180340 \approx 1\phi + 0$$

$$\phi^2 \approx 2.6180340 \approx 1\phi + 1$$

$$\phi^3 \approx 4.2360680 \approx 2\phi + 1$$

$$\phi^4 \approx 6.854120 \approx 3\phi + 2$$

$$\phi^5 \approx 11.090170 \approx 5\phi + 3$$

$$\phi^6 \approx \underline{\hspace{2cm}} \approx \underline{\hspace{1.5cm}}$$

$$\phi^7 \approx \underline{\hspace{2cm}} \approx \underline{\hspace{1.5cm}}$$

$$\phi^8 \approx \underline{\hspace{2cm}} \approx \underline{\hspace{1.5cm}}$$

1. Continue to find ratios F_{n+1}/F_n until you obtain an approximation for ϕ that is correct to six decimal places (1.618034). What is the least value of n that gives an approximation of ϕ correct to six decimal places?

2. In a geometric sequence, each term after the first term can be determined by multiplying the previous term by a constant. Normally a geometric sequence is represented by a_1, a_2, a_3, ... and r stands for the constant. In general, the terms of a geometric sequence can be described by the following formula, where a_n represents the nth term.

$$a_n = a_1 r^{n-1}$$

What are the values of a_1 and r in the sequence ϕ^1, ϕ^2, ϕ^3, ϕ^4, ... ?

3. Give an expression for the general term of the sequence ϕ^1, ϕ^2, ϕ^3, ϕ^4, ... as it would be described using the formula in Exercise 2. Then, use the formula to generate the first three terms of the sequence.

4. Use $(1 + \sqrt{5})/2$ for ϕ and the values of Fibonacci numbers F_1 and F_2 to show that $\phi^2 = (F_2)\phi + F_1$.

5. Use $(1 + \sqrt{5})/2$ for ϕ and the values of Fibonacci numbers F_2 and F_3 to show that $\phi^3 = (F_3)\phi + F_2$.

Binet's Formula:

$$F_n = \frac{1}{\sqrt{5}}\left(\frac{1 + \sqrt{5}}{2}\right)^n - \frac{1}{\sqrt{5}}\left(\frac{1 - \sqrt{5}}{2}\right)^n$$

Use a calculator and Binet's formula to find F_{10}.

Without the aid of a calculator, use Binet's formula to find F_2.

1. Without the aid of a calculator, use Binet's formula to find F_3.

2. Without the aid of a calculator, use Binet's formula to find F_4.

3. Use a calculator and Binet's formula to find F_{11}.

4. Use a calculator and Binet's formula to find F_{20}.

THE CHINESE TRIANGLE

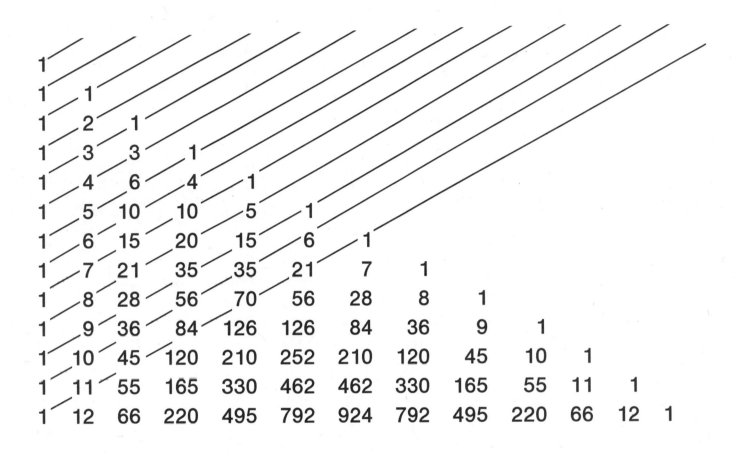

```
1
1    1
1    2    1
1    3    3    1
1    4    6    4    1
1    5   10   10    5    1
1    6   15   20   15    6    1
1    7   21   35   35   21    7    1
1    8   28   56   70   56   28    8    1
1    9   36   84  126  126   84   36    9    1
1   10   45  120  210  252  210  120   45   10    1
1   11   55  165  330  462  462  330  165   55   11    1
1   12   66  220  495  792  924  792  495  220   66   12    1
```

1. Expand each of the following binomials and write the resulting polynomials in the form of the Chinese Triangle. (Use 1's where you see no numerical coefficients.)

    a. $(a + b)^0 = $ ___

    b. $(a + b)^1 = $ ___ $a + $ ___ b

    c. $(a + b)^2 = $ ___ $a^2 + $ ___ $ab + $ ___ b^2

    d. $(a + b)^3 = $

    e. $(a + b)^4 = $

    f. $(a + b)^5 = $

2. After you have found all the expansions in Exercise 1, draw diagonals through them as was done with the Chinese Triangle in this lesson. Describe the sums of the binomial coefficients obtained by adding along your diagonals.

3. Find the sum of the numbers in each row of the Chinese Triangle for the first ten rows. Describe the sequence of sums.

MATHEMATICS AND MUSIC

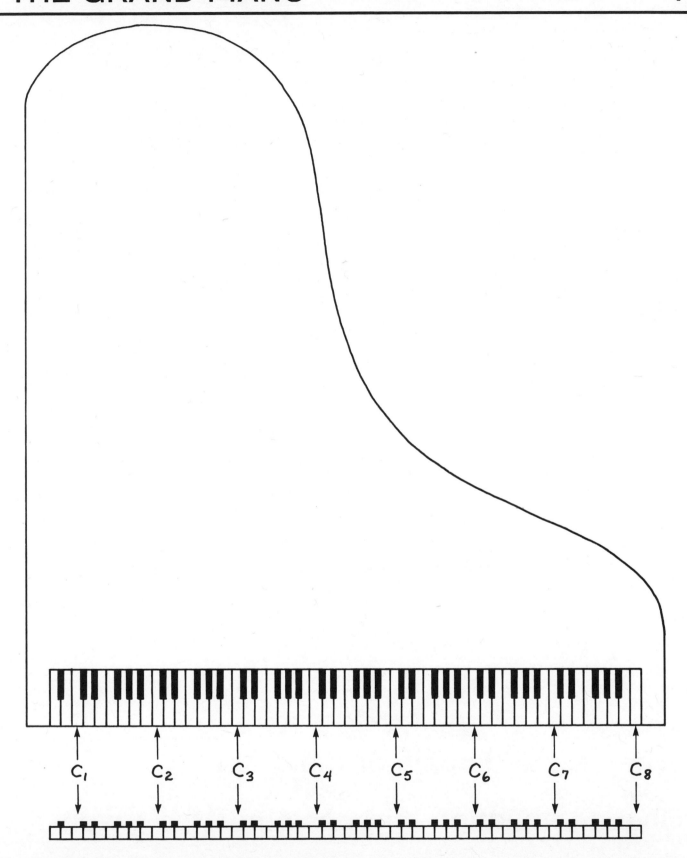

C_1 C_2 C_3 C_4 C_5 C_6 C_7 C_8

Graph the relative frequencies for C-notes of the piano.

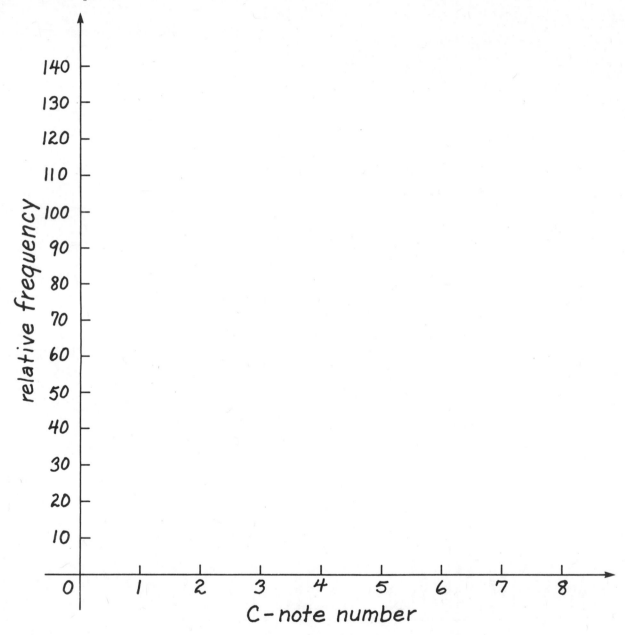

What is the domain of the graph?

Give an equation that describes the graph.

1. Graph $f(x) = 2^x$ where $x \in \{1, 2, 3, 4, 5\}$.

2. Graph $f(x) = 2^x$ where x is a real number.

3. Graph $f(x) = 3^x$ where x is a real number.

4. Graph $f(x) = 3^{-x}$ where x is a real number.

5. Graph $f(x) = 4(2^x)$ where x is a real number.

6. Graph $f(x) = 4(2^{-x})$ where x is a real number.

7. Graph $f(x) = 5(2^x)$ where x is a real number.

8. Graph $f(x) = 5(2^{-x})$ where x is a real number.

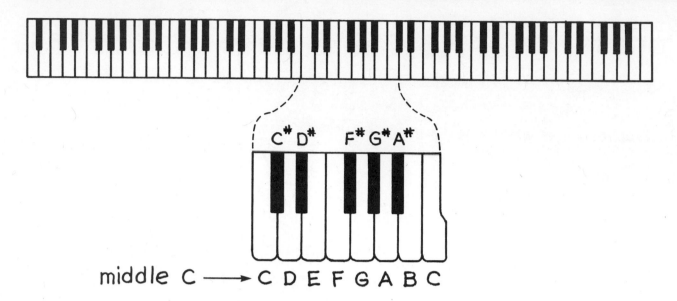

middle C ⟶ C D E F G A B C

Frequencies of consecutive notes are related by a common ratio, r.

Let f_n = frequency of nth note.

Let f_1 = frequency of middle C.

Write an equation relating f_n, f_1, and r.

Frequency of middle C is 261.6 cycles per second.
Frequency of next higher C is 523.2 cycles per second.
Find r.

Find the frequencies for all notes of the middle octave. Then complete the graph.

name	C	C#	D	D#	E	F	F#	G	G#	A	A#	B	C
n													
f_n													

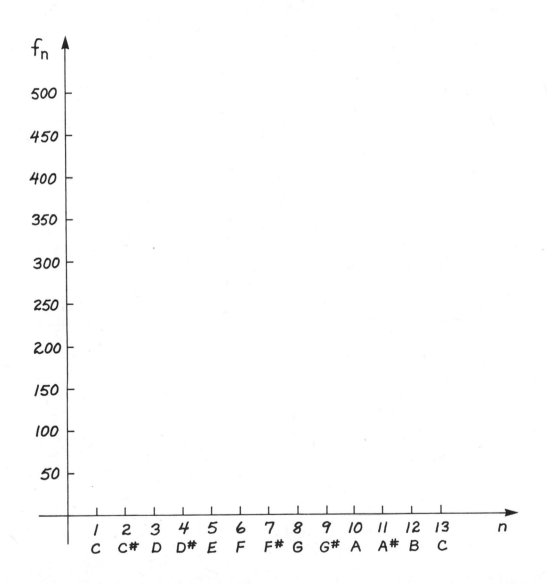

MORE PIANO RELATIONSHIPS

EXERCISES

1. Use the equation $f_n = f_1 r^{n-1}$ to verify that f_{13}, the frequency of C above middle C, is 523.2 cycles per second given that $f_1 = 261.6$ and $r \approx 1.05946$.

2. Describe how to extend the graph of this lesson to show the frequencies of notes in the next octave above the middle octave of a piano.

3. What is the frequency of F in the next octave above the middle octave of a piano?

4. What is the frequency of the upper C in the next octave above the middle octave of a piano? Describe *two* different methods for finding the frequency.

5. How are the frequencies of F in the middle octave of a piano and related to the frequencies of F in the next octave above?

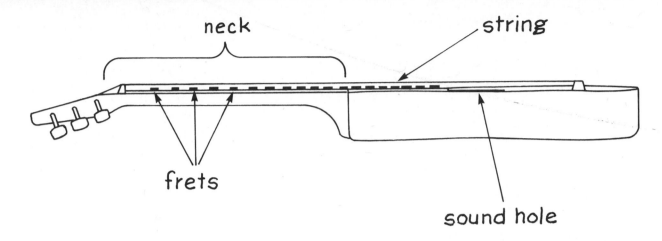

neck

string

frets

sound hole

Open E-string sounds E when plucked.	E-string pressed against first fret sounds F.	E-string pressed against second fret sounds F#.

Let *n* = position of fret on guitar

Let w_0 = wavelength of open E-string

Let w_n = wavelength of E-string sounding from *n*th fret

Write an equation that gives w_n in terms of w_0 and *n*. Use 1.0595 for $\sqrt[12]{2}$.

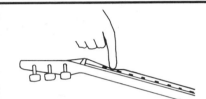

E-string

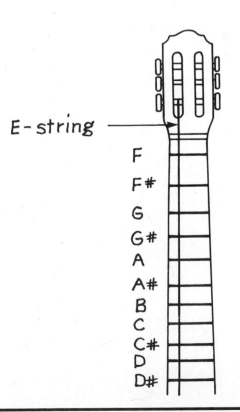

F
F#
G
G#
A
A#
B
C
C#
D
D#

Graph the wavelengths of the notes in a scale.
Sketch a smooth curve through the points.
Connect the guitar frets to the curve.
Where do they intersect the curve?

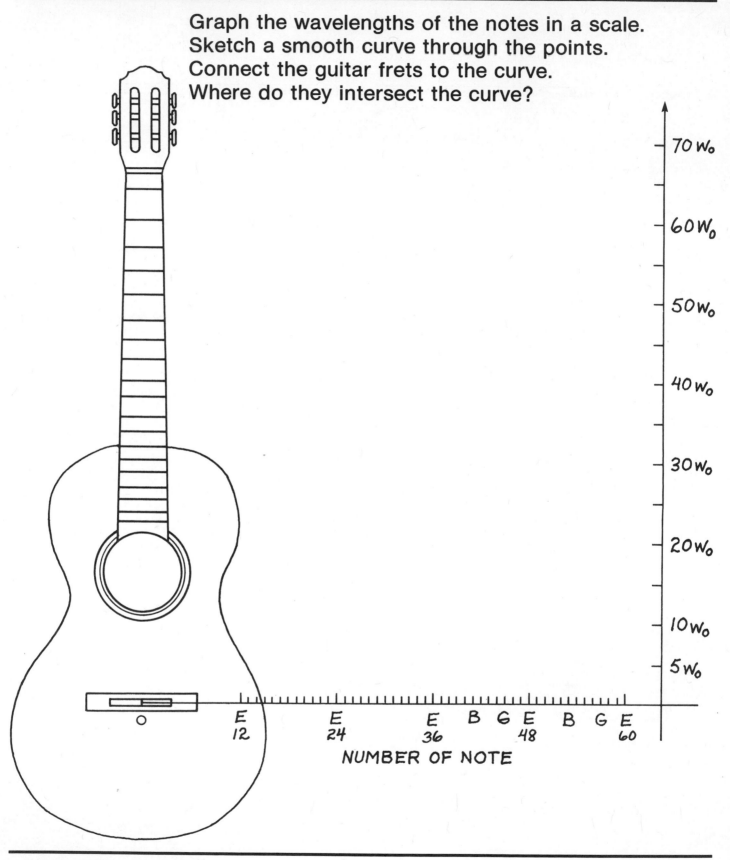

70 W₀

60 W₀

50 W₀

40 W₀

30 W₀

20 W₀

10 W₀

5 W₀

E
12

E
24

E
36

B G E
48

B G E
60

NUMBER OF NOTE

1. The equation $f_n \approx (261.6)(1.0595)^{n-1}$ where n is an integer from 1 to 13 approximates the frequencies of notes in the middle octave of a piano. Approximate the frequencies to the nearest tenth (or get them from your work in the lesson on Piano Relationships). Graph the frequencies on a line.

2. The equation $w_n \approx w_0 (1.0595)^n$ approximates the wavelengths of notes in the musical scale where w_0 stands for the wavelength of a high open E on the guitar and n stands for the number assigned to the frets of a guitar. Determine the relative wavelengths (values of $(1.0595)^n$) for the first 13 frets of a guitar. Graph the wavelengths on a line.

3. Compare the graph of Exercise 1 to the graph of Exercise 2.

4. In music, frequencies of notes and wavelengths of notes double over a full twelve-note octave. Within the octave there is an increase of about 5.95% for each successive note. In business, an investment will approximately double over a period of 12 years when the investment increases 5.95% each year. Write a formula to describe the business investment. Use n for number of years, P for the original amount of money invested, and A for the amount of money after n years. Compare your formula to the formula describing wavelengths of notes in a musical scale.

5. In general, for an investment with interest compounded annually, C dollars will approximately double in n years at p percent whenever p is an integral divisor of 72. This situation is often called the *Rule of 72*. Complete the following table based on the *Rule of 72*.

C dollars	p percent	n years	return on investment
100	6	12	200
100	8		200
100	3		200

6. According to the Rule of 72, at what rate of interest will an investment of $1000 double (approximately) when the interest is compounded annually and the money is invested for two years?

$$L = \frac{512}{p}$$

where L = length of pipe in feet

and p = pitch of pipe in cycles per second

Complete the table of values and sketch the graph.

p	16	32	48	64	80	96	112	128
L								

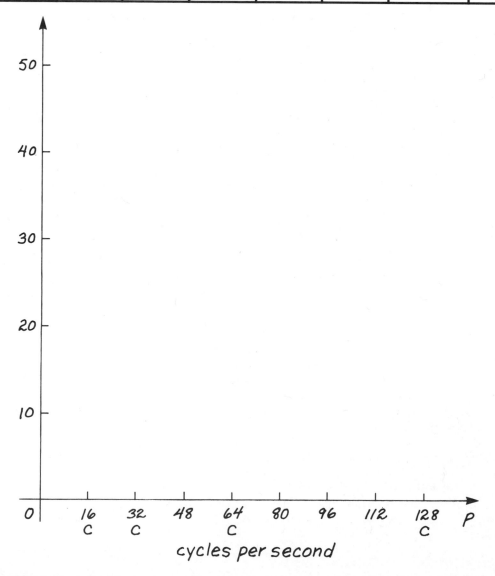

cycles per second

1. The graph of $L = 512/p$ is a hyperbola. In what two quadrants of the coordinate plane does the graph lie?

2. Why does the graph of $L = 512/p$ representing the relationship between the length of organ pipes to their pitch only lie in the first quadrant of the coordinate plane?

3. What is the pitch (in cycles per second) of an organ pipe that is 10.7 feet long?

4. What length of organ pipe will give a pitch of 96 cycles per second?

Complete the table of values. Graph the points you obtain and connect them with a smooth curve.

n	1	2	3	4	5	6	7	8	9	10	11	12	13	14	15	16	17	18	19	20
$(1.0595)^n$																				
$n(15°)$																				

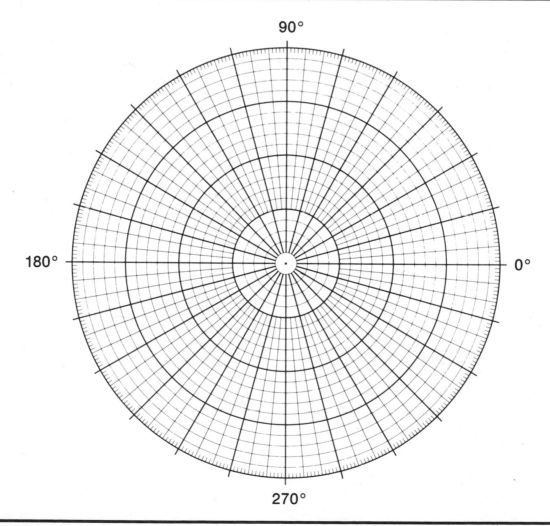

1. Draw golden rectangle *ABCD.*

2. Draw diagonal $\overline{DB}$.

3. Locate *E* on $\overline{DC}$ and *F* on $\overline{AB}$ so that figure *AFED* is a square. Notice that the small rectangle formed, rectangle *FBCE,* is a golden rectangle.

4. Draw diagonal $\overline{CF}$.

5. Locate *G* on $\overline{CB}$ and *H* on $\overline{EF}$ so that figure *ECGH* is a square.

6. Continue creating squares and golden rectangles as long as space allows.

7. Starting with point *F*, draw an arc of a circle with radius equal to the side of square *AFED.* The arc should start at *A* and end at *E.* Continue the process for *H, J,* and so on.

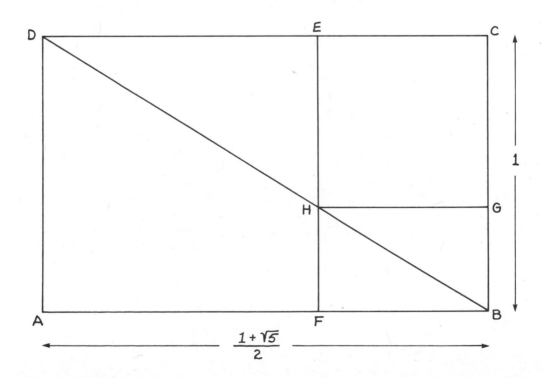

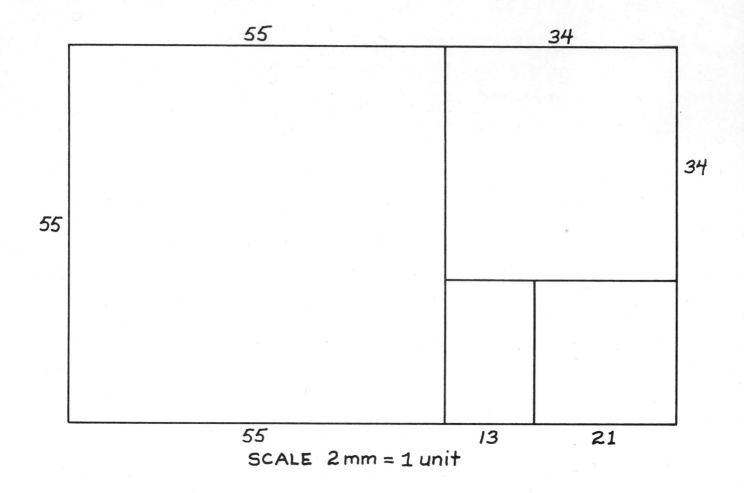

55

34

34

55

55   13   21

SCALE 2 mm = 1 unit

Continue the process of creating rectangles with successive Fibonacci measurements. Then draw quarter-circle arcs in each square.

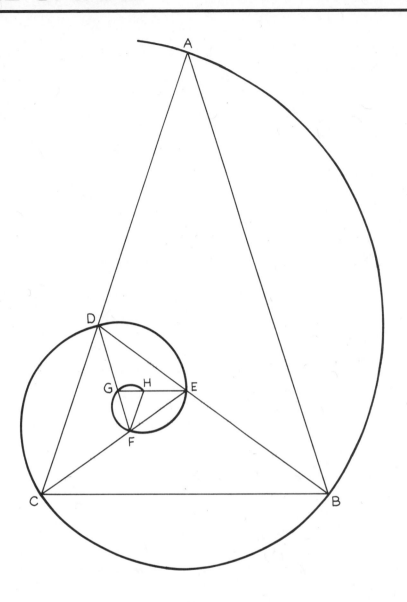

$GF = 1\phi$

$FE = 1\phi + 1$

$ED = 2\phi + 1$

$DC = 3\phi + 2$

$CB = 5\phi + 3$

$BA = 8\phi + 5$

What kind of triangles have been constructed?

_____

Logarithmic Spirals in the World

BOTANICAL

- spirals on a sunflower blossom (clockwise and counterclockwise)
- whorls of pineapple (clockwise and counterclockwise)
- spirals of pine cones (clockwise and counterclockwise)
- arrangements of florets in daisy blossoms
- graph of bacterial growth

ZOOLOGICAL

- horns of wild sheep
- sea shells (chambered nautilus, snails, and others)
- canary claws
- lion claw
- elephant tusks
- cochlea of human ear
- parrot beak
- teeth and fangs

ASTRONOMICAL

- galaxies of stars
- tail of comet curving away from sun

OTHER

- ocean waves
- shoreline of Cape Cod

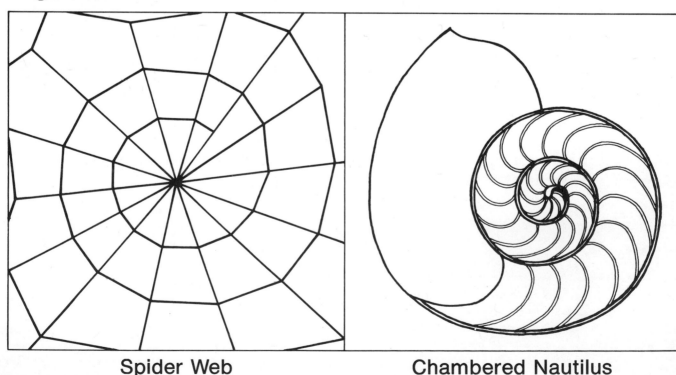

Spider Web   Chambered Nautilus

1. Find the polar coordinates of a point on the musical spiral for $n = 40$.

2. Find the polar coordinates of a point on the musical spiral for $n = 100$.

3. In theory, does the musical spiral ever end?

4. How can you extend the logarithmic spiral drawn in the golden rectangles to make it larger?

5. How can you extend the logarithmic spiral drawn in the "Fibonacci rectangles" to make it larger?

6. Can you continue the process described for Exercise 5 indefinitely?

7. If you continue the process described for Exercise 5, where would the very next square be added to the figure?

8. The general equation for a point (r, θ) in polar coordinates on a logarithmic spiral is $r = e^{\theta \cot K}$ where K is the measure of the angle that any radius makes with the spiral curve. Approximate K for the point $(1.0595, \pi/12)$.

Let $a_1, a_2, a_3, \ldots$ be radii of tangent circles defined as follows:

1. Every 4 consecutive circles are mutually tangent.

2. The radius of each circle is obtained by multiplying the radius of the preceding circle by the sum of the golden ratio and its square root.

$$a_n = a_1 \left(\phi + \sqrt{\phi} \right)^{n-1} \text{ where } n \text{ is a positive integer}$$

and a_n stands for the nth radius

Using 1.6180340 for ϕ gives the following approximation for the general term.

$$a_n \approx a_1 (2.890054)^{n-1}$$

Suppose $a_1 = 0.1$. Complete the following table.

term	a_1	a_2	a_3	a_4	a_5	a_6
approximate value	0.1					

Can you draw Coxeter's golden sequence of tangent circles?

Coxeter's Golden
Sequence of Tangent Circles

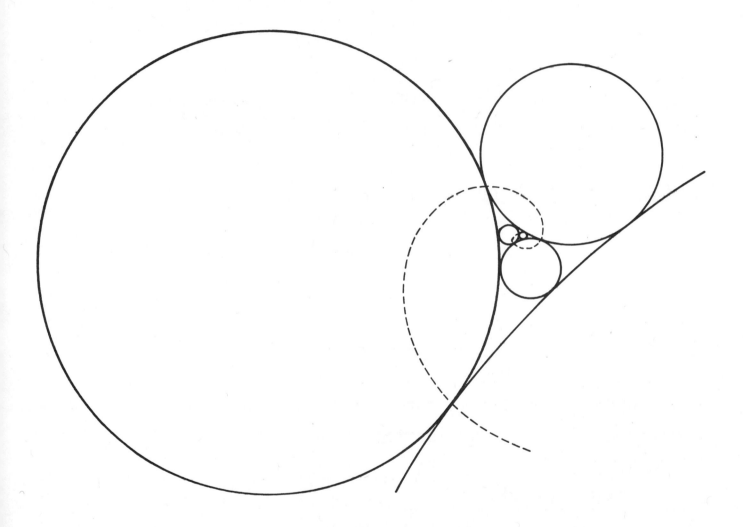

The contact points (points of tangency) lie on a logarithmic spiral.

1. Letting $a_1, a_2, a_3, \ldots$ stand for the radii of tangent circles in Coxeter's sequence, find values for a_6, a_7, and a_8 if $\phi = 1.618034$. Round your answers to the nearest three decimal places.

2. From the values you obtained in this lesson, calculate a_4/a_3.

3. Suppose you use 2.89 for the common ratio in Coxeter's sequence rather than 2.890054. In rounding to three decimal digits, which, if any, of the values for a_1, a_2, a_3, a_4, and a_5 differ?

REFERENCES

ONE   PROTECTING A COMPUTER'S SECRETS

Allman, William F. "Some Infinite Numbers are Pretty Big, Too." *Science '82,* Volume 8, No. 6 (June, 1983), p. 63.

Hellman, Martin E. "The Mathematics of Public-Key Cryptography." *Scientific American,* Vol. 242, No. 8 (August, 1979), pp. 146–157.

Kolata, Gina Bari. "Trial-and-Error Game That Puzzles Fast Computers," *Smithsonian,* (October, 1979), pp. 90–96.

TWO   PACKING PROBLEMS

Coxeter, H. S. M. "Close Packing of Equal Spheres," *Introduction to Geometry.* New York: John Wiley and Sons, Inc., 1961, pp. 405–407.

Gardner, Martin. "Mathematical Games: The Diverse Pleasures of Circles that are Tangent to One Another." *Scientific American,* Vol. 242, No. 1 (January, 1979), pp. 18–28.

Hoffer, Alan. *Teacher's Guide for Math in Nature Posters.* Palo Alto, CA: Creative Publications, Inc., 1978, p. 5.

THREE   A VERY SPECIAL NUMBER CALLED *e*

Arteaga, Robert F. *Building of the Arch,* 1967. Fourth Edition Printing, 1980. By Jefferson National Expansion Memorial Association.

Bakst, Aaron. *Mathematics, Its Magic and Mastery,* Second Edition. New York: D. Van Nostrand Co., Inc., 1952, pp. 313–328.

Boyer, Lee E. "Applications of Mathematics in the New Spirit of the St. Louis Arch." Lee E. Boyer, 1968.

Gardner, Martin. "Mathematical Games: The Imaginableness of Imaginary Numbers." *Scientific American,* Vol. 242, No. 8 (August, 1979), pp. 18–24.

Gardner, Martin. "Mathematical Games: In Some Patterns of Numbers or Words There May Be Less Than Meets the Eye." *Scientific American,* Vol. 242, No. 9 (September, 1979), pp. 22–25.

Manhard, Warren B., second. "Is Exponentiation Commutative?" *The Mathematics Teacher* (January, 1981), pp. 56–60.

FOUR   ALGEBRAIC FUNCTIONS

Burton, David M. *An Introduction to Abstract Mathematical Systems.* Reading, MA: Addison Wesley Publishing Co., Inc., 1965, pp. 21–25.

Deffeyes, Kenneth S. and Ian D. MacGregor. "World Uranium Resources." *Scientific American,* Vol. 243, No. 1 (January, 1980), p. 74.

Eaton, Bryan L. "Dots and Cubes." *The Mathematics Teacher* (February, 1974), pp. 161–164.

Griffith, Edward D. and Alan W. Clarke. "World Coal Production." *Scientific American,* Vol. 242, No. 1 (January, 1979), p. 46.

Kimura, Motoo. "The Neutral Theory of Molecular Evolution." *Scientific American,* Vol. 242, No. 11 (November, 1979), p. 104.

Likens, Gene E. et al. "Acid Rain." *Scientific American,* Vol. 242, No. 10 (October, 1979), p. 46.

Sawyer, W. W. "A Method of Discovery, I." *The Mathematics Student Journal,* Vol. 6, No. 1 (November, 1958).

Sawyer, W. W. "A Method of Discovery, II." *The Mathematics Student Journal,* Vol. 6, No. 2 (January, 1959).

Travers, K. J., Dalton, L. C., and Brunner, V. F. *Using Advanced Algebra,* Third Edition. River Forest, IL: Laidlaw Brothers Publishers, 1981, p. 308.

Weinreich, Gabriel W. "The Coupled Motions of Piano Strings." *Scientific American,* Vol. 242, No. 1 (January, 1979), p. 126.

FIVE   THE GOLDEN RATIO

Coxeter, H. S. M. *Introduction to Geometry.* New York: John Wiley and Sons, Inc., 1961, pp. 160–161, 166.

Huntley, H. E. *The Divine Proportion: A Study in Mathematical Beauty.* New York: Dover Publications, Inc., 1970, pp. 24–25, 28–30, 61, 140, 170–171.

Miller, Gordon. "Solutions." *The Point Subset.* Stevens Point, Wisconsin: The Department of Mathematics, University of Wisconsin—Stevens Point (September, 1980), p. 3. (Solution by Brian Yanny, Pius XI High School, Milwaukee, Wisconsin.)

Runion, Garth E. *The Golden Section and Related Curiosa.* Glenview, IL: Scott, Foresman and Company, 1972, pp. 21–25, 35–36, 52–53, 64–67.

Travers, K. J., Dalton, L. C., and Brunner, V. F. *Using Advanced Algebra,* Third Edition. River Forest, IL: Laidlaw Brothers Publishers, 1981, p. 465.

SIX   THE THIRTEENTH CENTURY STILL LIVES

Bergamini, David and the editors of Time-Life books. *Mathematics.* New York: Time, Inc., 1980, pp. 92–93.

Brousseau, Brother Alfred. "Fibonacci Sequences." *Topics for Mathematics Clubs.* Reston, VA: NCTM—Mu Alpha Theta, 1973, pp. 5–8.

Hoffer, William. "A Magic Ratio Recurs Throughout Art and Nature." *Smithsonian* (December, 1975), pp. 110–124.

Hoggatt, Verner E., Jr. *Fibonacci and Lucas Numbers*. Boston, MA: Houghton Mifflin Company, 1969, pp. 26–29, 48–50, 79–82.

Huntley, H. E. *The Divine Proportion: A Study in Mathematical Beauty*. New York: Dover Publications, Inc., 1970, pp. 49–50, 131–133, 145, 148, 161–165.

Land, Frank. "The Fibonacci Sequence and the Golden Section." *The Language of Mathematics*. Garden City, NY: Doubleday and Company, Inc., 1963, pp. 216–220.

Runion, Garth E. *The Golden Section and Related Curiosa*. Glenview, IL: Scott, Foresman and Company, 1972, pp. 71–73, 91–96.

Travers, K. J., Dalton, L. C., and Brunner, V. F. *Using Advanced Algebra,* Third Edition. River Forest, IL: Laidlaw Brothers Publishers, 1981, pp. 425, 429.

SEVEN MATHEMATICS AND MUSIC

Bakst, Aaron. *Mathematics, Its Magic and Mastery,* Second Edition. New York: D. Van Nostrand Co., Inc., 1952, pp. 295–296.

Gardner, Martin. "Mathematical Games: The Diverse Pleasures of Circles That Are Tangent to One Another." *Scientific American,* Vol. 242, No. 1 (January, 1979), pp. 24, 28.

Hoffer, Alan. *Teacher's Guide for Math in Nature Posters*. Palo Alto, CA: Creative Publications, Inc., 1978, p. 10.

Hoffer, William. "A Magic Ratio Recurs Throughout Art and Nature." *Smithsonian* (December, 1975), pp. 110–124.

Huntley, H. E. *The Divine Proportion: A Study in Mathematical Beauty*. New York: Dover Publications, Inc., 1970, pp. 100–102.

Jacobs, Harold R. *Mathematics: A Human Endeavor*. San Francisco: W. H. Freeman and Co., 1970, pp. 176, 271–273, 284–285, 290–291.

Land, Frank. "Logs, Pianos and Spirals." *The Language of Mathematics*. Garden City, NY: Doubleday and Company, Inc., 1963, pp. 128–132, 139–142.

Runion, Garth E. *The Golden Section and Related Curiosa*. Glenview, IL: Scott, Foresman and Company, 1972, pp. 54–55.

Stevens, Peter S. *Patterns in Nature*. Boston, MA: Little, Brown and Company, 1974, pp. 88–89.